# Solid Mechanics and Its Applications

Volume 209

*Series Editor*

G. M. L. Gladwell, Waterloo, Canada

For further volumes:
http://www.springer.com/series/6557

## Aims and Scope of the Series

The fundamental questions arising in mechanics are: *Why? How?* and How *much?* The aim of this series is to provide lucid accounts written by authoritative researchers giving vision and insight in answering these questions on the subject of mechanics as it relates to solids.

The scope of the series covers the entire spectrum of solid mechanics. Thus it includes the foundation of mechanics; variational formulations; computational mechanics; statics, kinematics and dynamics of rigid and elastic bodies: vibrations of solids and structures; dynamical systems and chaos; the theories of elasticity, plasticity and viscoelasticity; composite materials; rods, beams, shells and membranes; structural control and stability; soils, rocks and geomechanics; fracture; tribology; experimental mechanics; biomechanics and machine design.

The median level of presentation is the first year graduate student. Some texts are monographs defining the current state of the field; others are accessible to final year undergraduates; but essentially the emphasis is on readability and clarity.

C. M. Leech

# The Modelling and Analysis of the Mechanics of Ropes

Springer

C. M. Leech
Windsor House
Congleton, Cheshire
UK

ISSN 0925-0042 ISSN 2214-7764 (electronic)
ISBN 978-94-017-7638-7 ISBN 978-94-007-7841-2 (eBook)
DOI 10.1007/978-94-007-7841-2
Springer Dordrecht Heidelberg New York London


Softcover reprint of the hardcover 1st edition 2014


Printed on acid-free paper

Springer is part of Springer Science+Business Media (www.springer.com)

*To my wife, Brenda and to my daughters,
Andrea, Helen, Nicola and Suzanne*

# Preface

Linear fibre structures, typified by the many configurations of ropes, cables, stays and the smaller components, yarns and strands, have been the subject of analysis over many years. This is an attempt to formalise the subject and to emphasise the exploitation of these developments in the various engineering industries.

As well as considering the conventional theories of twisted assembly extension, the theories of loading of braided and plaited structures, bending, inter-component friction and component dilation and distortion are introduced. Also introduced is the modelling of wear or abrasion and of component heating.

Various modelling techniques have also evolved, typically the finite element theories and their associated finite element codes. These have been developed so that they form the regular tool in engineering offices, through their application to design and engineering assessment. However, they do use some element of continua even though they may account for many different components. Fibre structures on the other hand contain typically many million components, and it is not really appropriate to use such packages. This text considers these assemblies, the various sub-assemblies and their interaction behaviour. A notable inclusion is the friction between the components, and the various modes through which friction is applicable.

This text is solely focused on linear structures where the effect and consideration of the third dimension, length, is somewhat secondary to the primary dimensions, diameter and angular disposition. It is essentially a quasi-static investigation, although there is some small reference to general dynamics.

Finally there are other types of fibre assemblies, areas which include textiles and cloth, and volumetric assemblies which typically include bulk assemblies; these are not considered here.

## Acknowledgments

The author would like to thank the various colleagues at UMIST and TTI for various suggestions and inspiration that have surfaced over the years of contact. The author is also indebted to Mr. Gerry Needham for the preparation of various illustrations that are used in this text.

[illegible]

# Contents

# Chapter 1
# Introduction

**Abstract** This chapter, the introduces the modelling of linear fibre structure, ropes and cables firstly by outlining the structures and introducing the basics in the modelling approaches to be used.

## 1.1 Background

Fibres have been exploited in the construction of structures for many millennia both by man and by nature. In nature the integrity of plants, trees and indeed all life is dependent on the existence of fibre structures; these fibrous components could be grown within the structure as collagen is within the human body or they could be captured and processed by ingestion and digestion. Alternatively fibres have been harvested either deliberately or accidentally for processing into other structures typically for shelter and for protection as textiles (cloths). They have also been used in containment, tools and generally for the benefit of communities.

In the past these fibres have been grown naturally in grasses, flax, hemp, bamboo, harvested from animals in the form of for example bristles, and collected and extracted from wood; now these sources although still present are less important than the processing of fibres from non-fibres as exemplified by polymer and metal drawing and extrusion processes.

With the advent of manufactured fibres, there is significantly more control over their consistency, performance and uniformity both within the fibres and between batches; there is thus more confidence in their exploitation and the factors of safety (or ignorance) associated with their variability can be substantially reduced. These fibres are now used in configurations where the stresses are much larger than those that could be supported by natural fibres for the same fibre diameter or weight. Modern fibres and included here are the metallic and polymeric fibres, since their construction, shape and performance are designed, can be used in situations previously not considered for fibre exploitation.

C. M. Leech, *The Modelling and Analysis of the Mechanics of Ropes*,
Solid Mechanics and Its Applications 209, DOI: 10.1007/978-94-007-7841-2_1,

Because of the dependence on the fibres in new configurations and in order to reduce the safety factors associated with any structure, it is essential that the performance of these fibre structures be accurately determined. The use of computational facilities must rate highly in the assessment of such structures and consequently the analysis of fibre structures will require a substantial awareness of simple but detailed mathematics to establish the models, and various computational and algebraic skills to extract worthy engineering solutions.

At this point one should perhaps state what a fibre is; here a fibre is taken to be a component where one dimension is substantially larger than the other dimensions, typically the length is ten or more times larger than a diameter. This is a very casual prescription but more rigorous definitions required for fibre modelling will be established later. This definition includes of course filaments, conventional fibres and could indeed include various other linear structures such as yarns, strands, and ropes. These may be unusual inclusions within the definition of a fibre but a rope used in a suspension or cable stayed bridge indeed could for some analyses be considered a fibre. Also included at this point are beams and shafts although the bending and torsion stiffness are secondary to the axial stiffness; not excluded in the macroscopic sense are chains and other articulated components but since they are properly employed because the detail of their weight and discrete hinges are most important, they will not further considered.

## 1.2 Outline of Following Chapters

In Chap. 2, fibres and polymer fibre materials are discussed. Dimensionality is introduced as an essential component and the other assumptions then follow; axial stiffness and strength is a primary requirement and the secondary deformation modes, flexural and torsion are to a less extent important. The problems of measurement of linear structures leads to the classical textile measurements of 'lateral dimension' by way of weight and Tex or denier; these are carried through to the other material and component properties, typically as stiffness and strength.

Chapter 3 considers the general fibre paths and specifically helical geometry; the effect of deformation of the helix on the deformation of the constituent fibre component is analysed. Finally a comparison in the energy of extension, twist and flexure is made so that useful engineering approximations can be selected.

Having dealt with the fibre as the constituent for a structure, linear structures are then examined; these include yarns, cables and ropes. In Chap. 4, transversely continuous structures are considered; these are structure where the constituent components are so small and so numerous that it is appropriate to consider the structure as a continuum, although in the classis sense it would be anisotropic and inhomogeneous. In this chapter, the parallel assembly is also examined and the implications of variability in component length (or prestrain) are quantified.

In Chap. 5, the hierarchical nature of rope structures is identified for the analysis and modelling of ropes and cords and other similar structures; indeed it is indigenous in their manufacture and construction. The various terminology specific to the rope industry is detailed, and the process of structure normalisation is outlined.

Chapter 6 examines transversely discrete structures, those in which the components are easily countable and identifiable. The component (fibre, yarns, strands etc.) elements are assembled by twisting or weaving (braiding, plaiting) into another linear component whose dimensionality (L/d) is reduced from that of the constituent component. Tension and torsion behaviour are the focus, but by implication in the modelling transverse deformation is important is developing the model for the structure behaviour. This arises from the compression and distortion of components as the structure is loaded. Also in this chapter the various component nomenclatures are listed and the various component assembly configurations are described. The hierarchical nature of rope structures is important for the analysis and modelling of ropes and cords and other similar structures; indeed it is indigenous in their manufacture and construction. Finally bending of helical structures is introduced with the two extreme assumptions, no slip and zero friction.

Chapter 7 introduces contact forces and pressures and its modelling; consequent to this is friction as occurs within the structure and between the constituent components. Various friction or more correctly energy dissipative mechanisms are identified and are grouped into two categories; those that arise because of relative motion between components are labelled INTER modes and those that arise because of the deformation of the component are INTRA modes and dilation and distortion are in the latter category. Because of the inhomogeneous nature of the fibre structures, continuum theories do not necessarily lead to the best models. The quantification of dilation and distortion is achieved through the packing factor and the shape factor. Also included in this chapter is the modelling of structure life as limited by the continued abrasion between components. This uses the friction models and various wear criteria which must be measured by testing.

Chapter 8 considers the wear and heating of these structures, the prime considered cause of both being the repeated slip between the components due to deformation cycling of the structure. Classical heat measurement and transmission theories are reviewed, and considered for modelling of structure temperature rise. A global model for steady state is broached; heat in by abrasion equals heat out through the surface.

## 1.3 Closure

Finally the intention of this book is to detail the various analyses and mathematical models that can be used to quantify the behaviour of linear fibre structures; some are well established techniques and some are new and some evolved over many years. The majority of the models have been validated with experiments but some are conjectural and are so identified.

The application and exploitation of such models is dependent on various experimental measurements. For example, the component density and size is extensively used. This must be measured as also various other data, load-extension behaviour, friction and in the final section, wear and heat parameters. In many cases, it is neither possible nor convenient to measure these directly and in such situations similar, test or prototype structures are introduced. Then the relevant models are used to extract the various experimental data or coefficients which can then be applied to the other intended structures.

It has not been the intention to produce a text on the textile technology of ropes and cords as this would be better produced by others more suitably qualified; this text hopefully would serve those textile technologists as a tool to be used to reinforce their experimental research. It would also present the various models and associated model behaviour that can help in understanding the performance of linear fibre structures.

A caution that must be made here is the accuracy and appropriateness of the models; these have been constructed in a mathematical sense with precise and well defined components. This is not so in real life as fibre are not uniform nor homogeneous and the assembly into structures by a manufacturing process is not as controlled as the mathematical process of assembly. Thus the real final product must depend on its' constituents and their manipulation and if these are variable then there must be a divergence between measured and predicted performance. This can be illustrated by the assembly of a group of *identical fibres* into a parallel structure, discussed in Chap. 4; they do not behave in a group, as the sum of the individual components.

This does not relegate this text to the status of an abstract mathematical thesis. Its' function is three fold, first as a compendium of models and analyses to be used in the prediction of structural performance and in the assessment of the performance against standards, as a tool for determining trends and the *what if* concept of design and development and finally for the establishment of an analytical platform from which other mathematical arguments can be launched and other models developed.

# Chapter 2
# Fibre Geometry and Fibre Mechanics

**Abstract** Fibres and polymer fibre materials are discussed. Dimensionality is introduced as an essential component and the other assumptions then follow; axial stiffness and strength is a primary requirement and the secondary deformation modes, flexural and torsion are to a less extent important. The problems of measurement of linear structures leads to the classical textile measurements of 'lateral dimension' by way of weight and Tex or denier; these are carried through to the other material and component properties, typically as stiffness and strength. A detailed survey into the classification one dimensional material (fibre) properties such as viscoelasticity and anelasticity is conducted.

## 2.1 Introduction

In order to define a fibre it is necessary to specify some attributes; the first attribute must be dimensionality and to quantify this attribute the following characteristic dimensions are introduced, a length, $L$ and a transverse or lateral measurement typically diameter $d$ or in some cases, thickness $h$ and width $w$, Fig. 2.1.

The cross section is not necessarily circular, it can be elliptical, oval or indeed hollow or formed into rectangular or other defined shapes. The length dimension for fibre definition is assumed to be much larger than the transverse dimension, but it is not necessarily specified. In other words the fibres can be assumed to be infinitely long but there is a lower limit on length for a given diameter, or typically $L/d > 10$. The ratio ($L/d$) is sometimes called the aspect ratio. This limit on this (aspect) ratio is quite arbitrary but later will be more accurately quantified. Similarly the diameter is assumed to be small, and although it could be infinitesimally small compared with the length but it is limited in maximum size to say $d/L < 0.1$.

In the analysis of fibre structures, various assumptions are used; fibres are always assumed slender so that the lateral space occupied is insignificant. For structural analysis of fibre assemblies the assumptions relating to the fibre components are varied; some in detail relate to the stress distribution across the section namely constant or linear varying axial stress, or no shear deformation, and others

C. M. Leech, *The Modelling and Analysis of the Mechanics of Ropes*,
Solid Mechanics and Its Applications 209, DOI: 10.1007/978-94-007-7841-2_2,

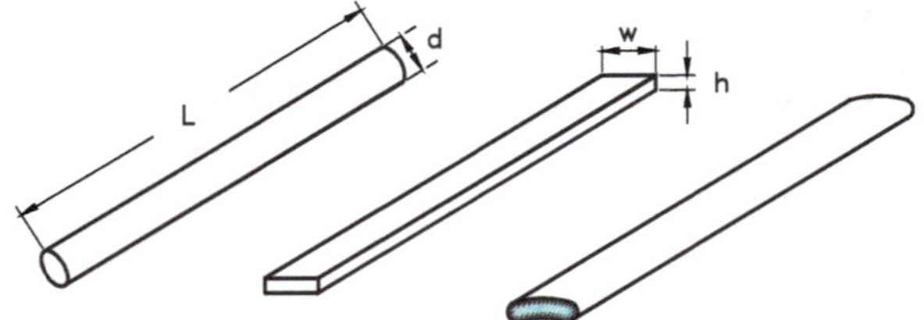


**Fig. 2.1** Typical fibres

macroscopic relate globally as a structural property as negligible flexural and/or torsion stiffness.

The aspect ratio above, a geometric concept is not be the sole specification of a fibre; they could also be specified according to their strength and flexibility so that there is a distinction between fibres and beams and shafts which also have a dominant geometrical specification ($L/d \gg 1$). Although the effects of flexure and twist will be considered it is advantageous to consider 'ab initio' the implications of these mechanisms and of the resistance to these deformations.

The assumption of a fibre is considered first for a fibre behaving as a structure and secondly for the fibre behaviour within a structure.

## 2.2 Fibre Dimensionality

As outlined above, dimensionality can be used to define a fibre and to do this a geometric parameter or measure $G$ is introduced; a geometric length $L_g$ is specified such that $L_g/d = G \gg 1$ and if the component length $L$ is larger than $L_g$ then the component can be assumed to be a fibre.

However if the structure contains fibres, as in ropes and cloth, there is a size quantity that specifies the fibre effect; if the characteristic width or diameter dimension of the structure is $D$ then the constructional dimensionality is $d/D$. If this is small (<0.001) as in typical synthetic fibre ropes where there could be 1 million fibres, then the constituents could be considered fibres; if however this ratio is larger than 0.001, the thickness effect of these become significant as in wire ropes where $d/D$ is typically larger than 0.05, from a count of 400 wires.

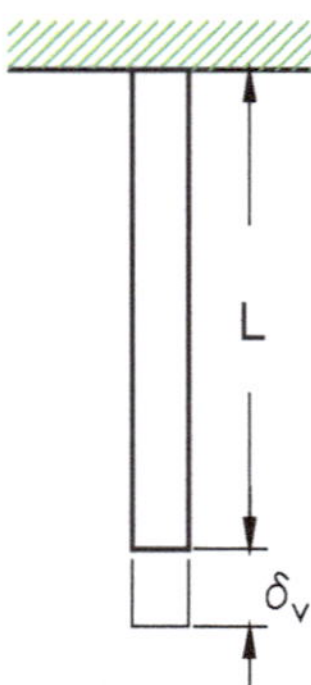

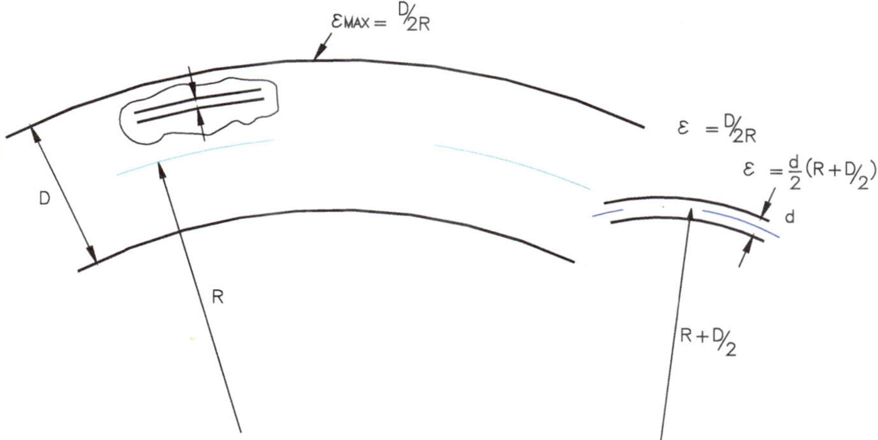


**Fig. 2.2** Bending of a fibre showing strain and curvature

As an illustration, consider a structure with a characteristic diameter $D$ in bending, Fig. 2.2; the typical maximum strain due to bending $\varepsilon_{max}$ occurs at the outside and inside edges and if the radius of curvature of the structure due to bending is $R(>D)$ then this maximum strain is $\pm D/2R$. A fibre at this position in the structure has a radius of curvature $R \pm D/2$ and whereas the fibre extension strain $\varepsilon_s$ is $D/2R$, the accompanying bending strain $\varepsilon_b$ is $d/(2R \pm D)$. It thus follows that the effort in bending the fibre is insignificant compared to that in stretching it and the bending resistance of the fibre is justifiably ignored in this configuration.

A similar criterion could be applied to area and volume structures in which the characteristic transverse dimension $D$ of the structure is compared to the fibre diameter d; large ratios of $D/d$ suggest that only fibre length and axial strength are important, whereas small ratios suggest that fibre flexure and twist are significant, Fig. 2.3.

In the former case the structure is likely to be dense, the fibre count being used to 'fill' the structure and in the latter case, there are likely to be gaps or voids resulting in a lower packing ratio. This, the packing ratio can be defined here as the

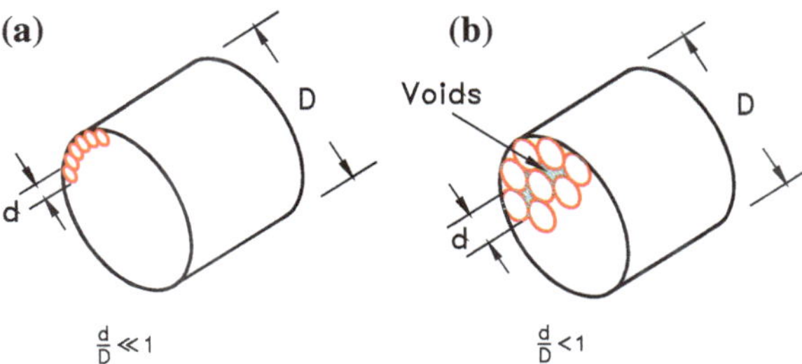


**Fig. 2.3** The assembly of fibres into a structure showing voids

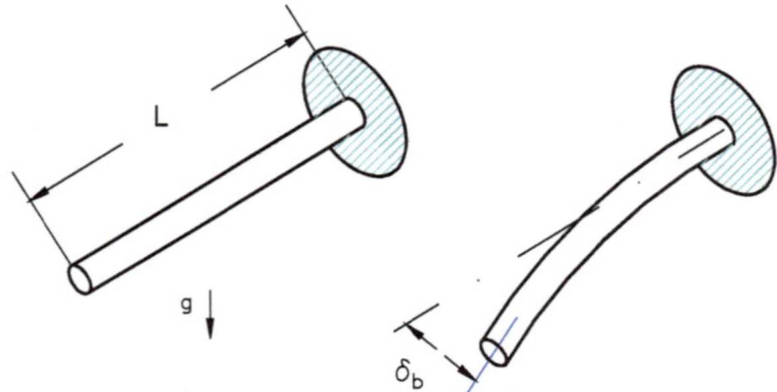


**Fig. 2.4** Flexure of stiff and limp fibres

ratio of the accumulated projected area of all the constituent fibres to the total projected area of the structure.

The simplistic criterion for a fibre is one in which extensional stiffness dominates and bending stiffness is negligible. To see this, consider two fibres of the same length cantilevered in a gravity field; if the two fibres are different in material or diameter, one will bend more than the other. That with the less bend could be identified more as a bristle whereas the other with larger sag will be classed as a limp fibre, Fig. 2.4.

In order to compare the effects of loading a fibre, two characteristic lengths are identified, Fig. 2.5. The first length $L_v$ is for a fibre hanging vertically in a gravity field. The extension or inline deflection for a given fibre length $L$ is given as follows,

$$\frac{\delta_v}{L} = \frac{\rho g L}{2E}$$

and the strain energy of the fibre, length $L$ is

$$U_v = (\rho g)^2 \frac{AL^3}{6E}.$$

The limiting fibre length $L_v$, or that length which can be suspended without exceeding the a critical strain $\varepsilon_c$ is $\frac{L_v}{d} = \frac{E\varepsilon_c}{\rho g d}$. The second characteristic length $L$, is determined for a fibres supported horizontally as a cantilever beam; the tip

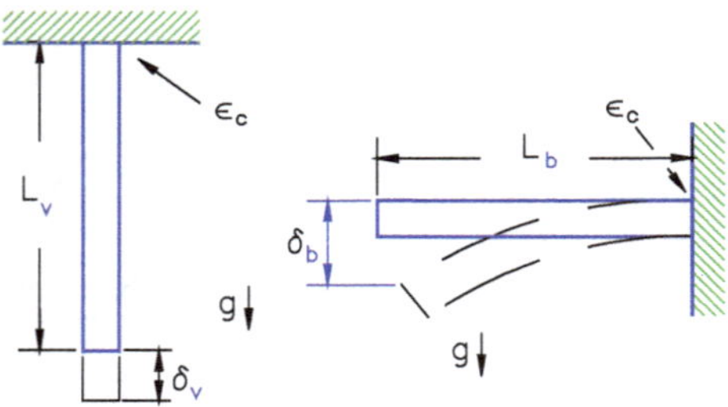


**Fig. 2.5** Fibre characteristic lengths

deflection is $\frac{\delta_b}{L} = \frac{\rho g L}{8E_S^2}\left(\frac{L}{d}\right)^2$, where $A(sd)^2$ is the second moment of area of the cross section about the bending axis and where $s$ is a shape factor and is 0.25 for a circular section. The strain energy in bending is

$$U_b = (\rho g)^2 \frac{AL^3}{20E_S^2}\left(\frac{L}{d}\right)^2.$$

The limiting cantilever length $L_b$, the maximum length without exceeding the critical strain is $\frac{L_b}{d} = 2s\sqrt{\frac{E\varepsilon_c}{\rho g d}}$. The bending length and hanging length now can be related, $\frac{L_b}{d} = 2s\sqrt{\frac{L_v}{d}}$.

The following table gives the characteristic lengths for a 1 mm diameter steel wire and a 1 mm diameter Kevlar fibre; the wire diameter is fairly typical whereas Kevlar fibres are usually much thinner.

| Material | Steel | Kevlar |
|---|---|---|
| Modulus (GN/m$^2$) | 200 | 100 |
| Density (kg/m$^3$) | 8,000 | 1,440 |
| Yield strain | 0.001 | 0.04 |
| Hanging length (m), $L_v$ | 2,500 | 283,000 |
| Flexure length (m), $L_b$ | 0.79 | 8.4 |
| $L_v/L_b$ | 3,164 | 33,690 |

For a heavy weak thick fibre, the $L_v/d$ is small and $L_b/d$ is relatively large; a bending slenderness criterion now dominates, and the component is classed as a stiff fibre or beam. However for a light strong thin fibre, the $L_v/d$ is large and $L_b/d$ is relatively small and the extension slenderness criterion dominates, classed as a flexible fibre where the structural resistance to flexure is negligible.

The implementation of these criteria will discriminate between the tendency to flex or stretch; for a given fibre with a characteristic length $L$ and hence a geometric slenderness, the assumptions relating to the analysis of fibre structural behaviour will be decided by the ratios $L_b/L$ and $L_v/L$.

## 2.3 Fibre Properties

The first property to be quantified and possibly the least significant is the length $L$; if it exceeds the geometric length $L_g$ then it is classified as a fibre. The end effects due to the proximity of presenting an exposed cross section are local and indeed important for short fibres and is the subject of much investigation in the use of short fibres in composites and non-woven textiles [1, 2].

### 2.3.1 Fibre Size

The lateral or transverse dimension $d$ is of course very important since the fibre strength, elasticity, weight are approximately proportional to $d^2$ and the flexibility to $d^4$ respectively. This dimension quantifies the amount of material used within the fibre and its effectiveness. Because of its importance, its measurement is critical; conventional linear measurements of this very small quantity are not very accurate, and indeed if the section is neither conventional nor circular then the measurement produced is not sensible. The conventional (textile) measurement is linear density or weight/unit length since both of these can be measured conveniently; the ovalness, ellipticity or noncircularity of the section are accounted for in the implicit averaging process of sampling a finite length.

### 2.3.2 Fibre Weight

The linear density $\omega$ simply related to the material density $\rho$ and the section area $A$ by $\omega = \rho g A$.

The effective or equivalent diameter $d$ is given by the following, $d = 2\sqrt{\omega/\pi\rho g}$, and the cross section area $A$ by $A = \omega/\rho g$.

The units associated with linear density are lb/ft, kg/m, denier (gm/9,000 m), Tex (gm/km) and dtex or decitex (gm/10 km). The 'size' could be measured using the linear density or could be the diameter or circumference.

### 2.3.3 Fibre Material Stress–Strain Behaviour

The strength of a fibre is the maximum load that it can carry; however there are variations to this definition [3, 4]. For an elastic structure, there is a load above which the fibre is no longer elastic, and this is the yield point. Before proceeding further, the following definitions are required:

(a) An *elastic* material, Fig. 2.6a, is one in which there is a one-to-one correspondence between stress and strain, and there is a stress-free strain-free state; this definition excludes viscoelastic materials since for any value of strain the accompanying stress depends also on the strain rate. Any material that is not elastic is said to be *anelastic.*
(b) A *plastic* material, Fig. 2.6b only has a stress-free strain-free state prior to any loading; once it has been loaded beyond the yield point, there is no correspondence between stress and strain and the behaviour of the material is governed by its state of strain, the history of deformation which resulted in that state, and the strain rate.

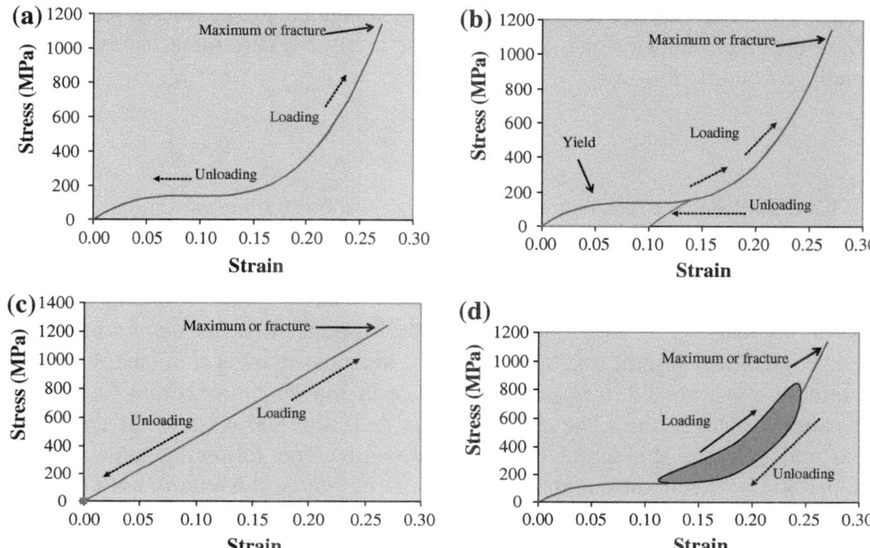


**Fig. 2.6** **a** Nonlinear elastic material behaviour. **b** Nonlinear elastic/plastic material behaviour. **c** Linear elastic material behaviour. **d** Nonlinear viscoelastic material behaviour

The *yield point* is a point in the stress strain space, that defines the limit of elasticity. If the yield stress or strain is exceeded the material is no longer elastic and is either plastic or has fractured.

(c) A *linear elastic* material, Fig. 2.6c, is an elastic material in which the stress is proportional to the strain; in other words the stress strain correspondence is linear.

(d) A v*iscoelastic* material is an anelastic material with elastic characteristics and a strain rate dependency, Fig. 2.6d.

Since fibres can support axial and shear loads, torque and bending moments, and the fibre material experiences stress, there is a limiting connection between component load and material stress. The conventional definition of stress is the limit of load over the carrying area as that area is taken to the limit, as follows

$$\sigma = lim \frac{\delta F}{\delta A}$$
$$\delta A \rightarrow 0$$

where $\sigma$ is the stress, units are typically $\text{lbin}^{-2}$ (psi), or $\text{Nm}^{-2}$ (Pa). This definition using the fibre axial load $F$, gives the fibre material direct stress, based upon area,

$$\sigma = \frac{F}{A}.$$

However if in the above the fibre area is relegated to secondary importance behind the linear density, the stress could be and for textile materials is conventionally written as follows

$$\sigma = \frac{F}{\rho g A} = \frac{F}{\omega}.$$

This definition of stress is the *specific stress* and has the measurements of force/linear density with units, lb/lb per ft or ft, or N/tex or N/denier or km; the length unit is that length of fibre which when vertically suspended in air gives that specific stress at the suspension. This length will vary with surrounding medium since when hanging in, for example water the suspension stress will be less since there is a reduced weight due to buoyancy. This measurement unit, specific stress is common for fibre and linear structures since in enables direct comparisons to be made between materials. The specific stress definition carries over of course, to yield and fracture stress and to elastic modulus. The following table gives a comparison between various materials in different units; the data is not definitive and is quoted for comparative purposes both for the difference arising from use of various materials and for representation in different units. It is useful to compare the neighbouring columns, steel and Kevlar from the point of strength; if conventional or engineering stress (based upon cross section area) is used the two material are comparable. However if strength is compared by amount (weight) of material used then the polymer materials are favoured. This is significant in any cable configuration since part of the strength capability must be used in supporting the cable structure and the lighter materials are preferred; this differential is accentuated in denser than air media such as water since this has the effect of subtracting unity from the density. For Kevlar/steel comparisons the strength gain for the same weight of material is steel 7.35 in air but is 24 in water.

Figure 2.7 illustrates graphically the approximate stress–strain behaviour of various materials used in fibres.

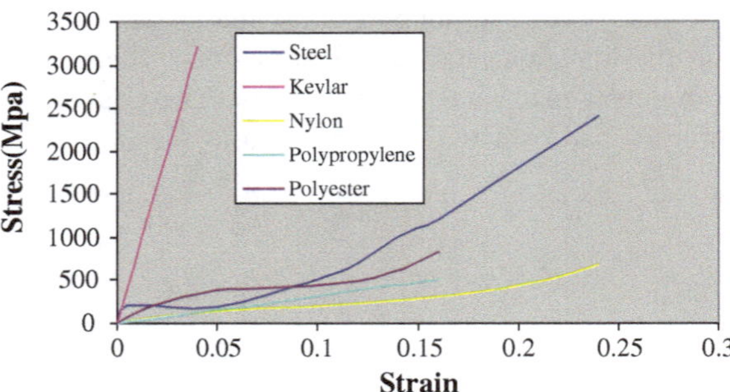


**Fig. 2.7** Comparative material stress behaviour

The following table summarises typical material and fibre data

| Material | Steel | Kevlar | Nylon | Polypropylene | Polyester |
|---|---|---|---|---|---|
| Diameter (mm) | 1 | 0.0122 | 0.0499 | 0.0304 | 0.0227 |
| Density (gm/cm$^3$) | 8 | 1.44 | 1.14 | 0.92 | 1.38 |
| *Linear density* | | | | | |
| gm/km, tex | 6,283 | 0.167 | 2.23 | 0.668 | 0.559 |
| Denier | 56,434 | 1.5 | 20.06 | 6.01 | 5.026 |
| *Modulus* | | | | | |
| GN/m$^2$ | 200 | 100 | 4.6 | 2.5 | 14 |
| N/tex | 25 | 69 | 4.04 | 2.72 | 10.14 |
| gm/denier | 1,140 | 3,146 | 182.8 | 123 | 459.6 |
| Km | 2,564 | 7,078 | 411 | 277 | 1,034 |
| *Breaking stress* | | | | | |
| MN/m$^2$ | 2,400 | 3,200 | 680 | 500 | 830 |
| N/tex | 0.3 | 2.22 | 0.6 | 0.54 | 0.6 |
| gm/denier | 3.4 | 25 | 6.8 | 6.2 | 6.8 |
| Km | 30.5 | 226 | 61 | 55.4 | 61.3 |
| *Breaking limit* | | | | | |
| Breaking strain | 0.24 | 0.04 | 0.1–0.3 | 0.15–0.4 | 0.16 |

## 2.4 Primary Deformation, Extension

The primary action of fibres in structures is to resist axial deformation by developing tensile forces in the fibres. In the development of fibres, either by manufacture or by growth, the consistency of material properties across the fibre cross section is varied or this is due to the difference in exposure of the outside face or skin. Consequently the stress distribution across the section would be nonuniform, probably larger in the interior for a simply tensioned fibre. However because of the large aspect ratio and smallness of the diameter, this nonuniformity is ignored since there is a small dimension over which the stress can vary. The variation of stress across the section will be significant in the secondary deformations since twist and bending can only be supported by a variation in stress and this will be considered later in this chapter.

For primary loading (axial tensioning) it will be assumed that the stress is constant over the cross section; such an assumption will also be used when the fibre are loaded by shear stresses caused by the interaction of fibres imbedded in composite matrices. In this case there must be a variation of stress along the fibre length but it is assumed that there is no variation across the fibre. The behaviour of an elastic fibre under axial tension can be written as a function, $\sigma = \Im(\varepsilon)$ where $\varepsilon$ is the fibre strain, $\sigma$ is the stress, Fig. 2.8a.

The above equation denotes a one-to-one correspondence between stress and strain, a requirement for the assumption of elasticity. The second requirement, a stress free, strain free state is assured by requiring that $\Im(0) = 0$.

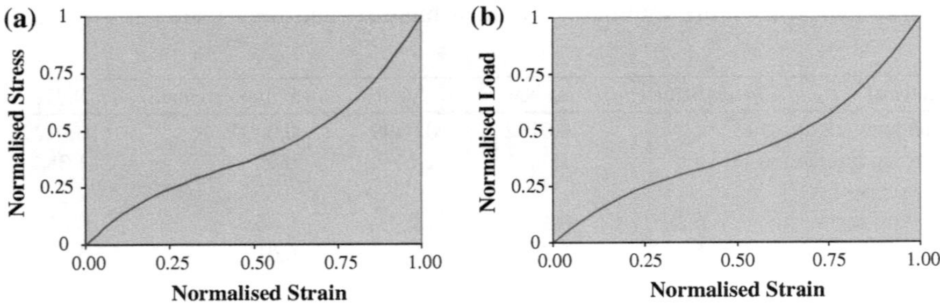


**Fig. 2.8** **a** Normalised stress strain behaviour. **b** Normalised load behaviour

To unify the above function bearing in mind the scope of materials available for use in fibres, the following is used

$$\frac{\sigma}{\sigma_f} = \Im(\eta)$$

where $\sigma_f$ is the fracture or maximum stress and $\eta$ is the normalised strain ($=\varepsilon/\varepsilon_f$); the function $\Im$ is now forced so that $\Im(1) = 1$, Fig. 2.8b. Since the stress has been normalised by the fracture stress, the above equation can be used for engineering or specific stresses and indeed in the context of fibre behaviour, the stress can be replaced by fibre tension and fibre breaking load.

## 2.4.1 Polynomial Approximations

Specifically, using a polynomial fit for the function $\Im$, the above can be written as $\sigma = \sigma_f \sum_{i=1}^{i=n} a_i \eta^i$ where $\sum_{i=1}^{i=n} a_i = 1$. A least square fit can be used to estimate the coefficients $a_i$, but the constraints $\Im(0) = 0$ and $\Im(1) = 1$ must be implemented; that is $a_0 = 0$ and $a_n = \sum_{i=1}^{i=n-1} a_i$. If the experimental data set comprises of $m$ data pairs, $\{\sigma_j$ and $\eta_j\}$, the error induced at each data station from using the polynomial fit is $\varepsilon_j = \sigma_j - \sum_{i=1}^{i=n-1} a_i \eta_j^{\,i} - \left(1 - \sum_{i=1}^{i=n-1} a_i\right) \eta_j^{\,n}$ and the implementation of the method of least squares requires that $\frac{\partial \sum_{j=1}^{j=m} (\varepsilon_j)^2}{\partial_{a_i}} = 0 \; for \; i = 1, 2, \ldots, n-1$.

There are $n-1$ equations for the $n-1$ unknown $a$'s, that is for $a_1, a_2, \ldots, a_{n-1}$. Although the coefficients $a_1, \ldots, a_n$ could be established using the least squares polynomial fit, a fit could also be achieved using the measured fibre or material properties. For example if the modulus at zero strain $E_0$, the modulus at fracture strain $E_f$, and the energy to fracture $W_f$ are measured, a cubic polynomial can be fitted as follows

$$\frac{\sigma}{\sigma_f} = \frac{E_0\varepsilon_f}{\sigma_f}\eta(2-5\eta)(1-\eta)^2 + \frac{1}{2}\frac{E_f\varepsilon_f}{\sigma_f}\eta^2(3-5\eta)(1-\eta) + 30\frac{W_f}{\sigma_f\varepsilon_f}\eta^2(1-\eta)^2 - \eta^2(6-5\eta)(2-3\eta)$$

## 2.4.2 Strain Energy

There exists a unique strain energy density function $W_d(\varepsilon)$ for elastic materials, Fig. 2.9a, and for linear structures such as fibres the strain energy density function is

$$W_d = \int_{\varepsilon=0}^{\varepsilon} \sigma \, d\varepsilon$$

Employing the above polynomial expression for stress, the strain energy density function is

$$W_d = \sigma_f\varepsilon_f \sum_{i=1}^{i=n} \frac{a_i\eta^i}{i+1}.$$

The units associated with the strain energy density $W_d$ are J/m$^3$ when engineering stress (N/m$^2$) is used, kJ/gm when specific stress (N/tex) is used, and J/m when the fibre tension is used; each representation is useful but the last is especially significant since for a given length of fibre it will yield the its limiting energy content. This strain energy function and the above stress function are functions of axial strain only since it is assumed that the fibre axial stress or tension is not a

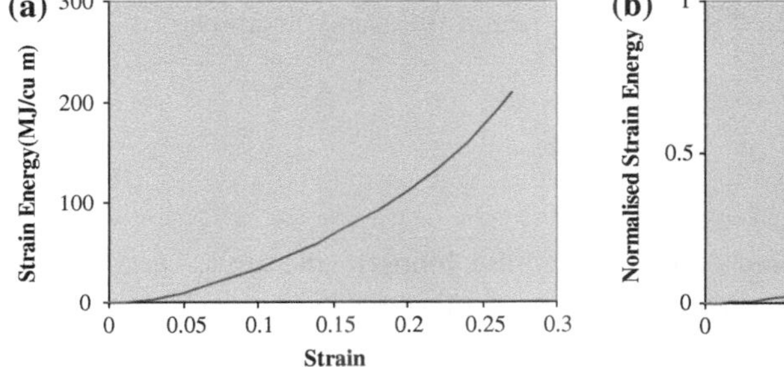


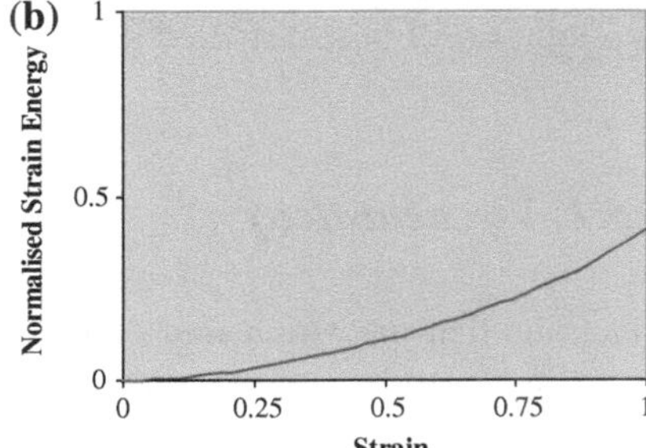


**Fig. 2.9** **a** Strain energy. **b** Normalised strain energy

function of fibre twist or flexure and these are *simple* fibres. Fibres that have a coupling between axial stress and twist and/or flexure are *nonsimple* fibres and will be considered later under secondary deformations.

The advantage of using normalised stress as a function of the normalised strain is emphasised; as a consequence, all fibre behaviour is characterised by the fracture stress, fracture strain and the function $\Im(\eta)$, Fig. 2.9b. This latter is limited to the unit square and the energy required to result in a specific strain is the area under the curve. The energy to fracture, $W_f$ is typically $\kappa\sigma_f\,\varepsilon_f$, where the factor $\kappa$ is typically between 0.25 and 0.75 and the normalised strain energy is $W/\sigma_f\,\varepsilon_f$.

Finally the connection between stress (or specific stress or tension) and strain energy or the work done in achieving a specific state of strain is as follows

$$\sigma = \frac{\partial W_d}{\partial \varepsilon},\ W_d = \int_{state1}^{state2} \sigma\, d\varepsilon.$$

## 2.5 Anelasticity

Any material that are not elastic are said to be anelastic. The necessary conditions for elasticity are a one to one correspondence between stress and strain and the existence of a simultaneous stress-free strain-free state. Two variants are considered here, those materials that are strain rate dependent and those that that exhibit hysteretic properties.

Restricting the development to fibre behaviour, the material stress will be synonymous with and written as fibre load, as there is a simple proportionality between the two.

The fibre load function can be written

$$F = \Im\left(\varepsilon, \frac{d\varepsilon}{dt}, H(\varepsilon_p)\right)$$

where $d\varepsilon/dt$, $\dot{\varepsilon}$ is the strain rate and $H(\varepsilon_p)$ is the hysteresis associated with the strain history. $\varepsilon_p$ is the post yield strain similar to that used in plasticity of materials. The two topics will be considered sequentially in the following sections.

### 2.5.1 *Viscoelasticity*

Materials that are strain rate dependent exhibit loops in the stress–strain (load-extension) plane when the material is cycled about a mean strain; the opening in these loops is governed by material viscoelastic parameters and by the cycle parameters, specifically the speed or frequency of the cycles and the strain amplitude. For linear viscoelastic materials the theory is well established whereas for nonlinear viscoelasticity the developments are not formalised. The linear theory will be reproduced and extended to include nonlinearity.

First consider a general functional representation of a fibre material that only responds to a single strain, and in this case is the fibre extensional strain $\varepsilon$,

Consider the strain rate effect, viscoelasticity, written as

$$F = \Im\left(\varepsilon, \frac{d\varepsilon}{dt}\right)$$

Application of a Taylor approximation to this function gives

$$F = \Im\left(\varepsilon, \frac{d\varepsilon}{dt}\right) \approx \Im(\varepsilon, 0) + \frac{d\varepsilon}{dt}\left|\frac{\partial \Im\left(\varepsilon, \frac{d\varepsilon}{dt}\right)}{\partial\left(\frac{d\varepsilon}{dt}\right)}\right|_{\frac{d\varepsilon}{dt}=0}$$

This approximation, the first two terms in the Taylor approximation results in a model for materials that are nonlinearly elastic and linearly viscoelastic, since the viscoelastic effect is proportional to the strain rate. In fact this is the more general form of the Kelvin-Voigt model, composed of a spring element in parallel with a dashpot element, Fig. 2.10.

This is generally applicable for materials subject to low frequency straining; for large strain rates more than two terms in the Taylor approximation are required and will lead to other more complex representations. Some high strain rate approximations are included later in this section.

## 2.5.2 Linear Viscoelasticity

Recalling the Taylor approximation

$$F = \Im\left(\varepsilon, \frac{d\varepsilon}{dt}\right) \approx \Im(\varepsilon, 0) + \frac{d\varepsilon}{dt}\left|\frac{\partial \Im\left(\varepsilon, \frac{d\varepsilon}{dt}\right)}{\partial\left(\frac{d\varepsilon}{dt}\right)}\right|_{\frac{d\varepsilon}{dt}=0}$$

The first term is strain dependent only and using the previous normalised formula $F = F_f f(\eta)$ where $F$ is fibre load, $F_f$ is fibre fracture load, $\varepsilon_f$ is fibre fracture strain and where $\eta = \varepsilon/\varepsilon_f$, and is bounded $0 \leq \eta \leq 1$.

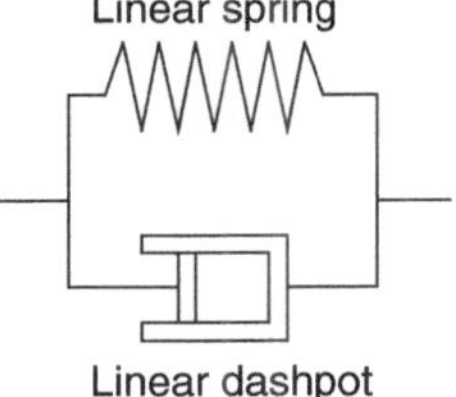


**Fig. 2.10** The linear Kelvin-Voigt viscoelastic model

For linear elasticity, $F = F_f\eta = M\varepsilon$ where $M$ is the modulus $\left(= F_f/\varepsilon_f\right)$ and for linear viscoelasticity

$$F = F_f\left(\eta + \lambda\frac{d\eta}{dt}\right) = \frac{F_f}{\varepsilon_f}\left(\varepsilon + \lambda\frac{d\varepsilon}{dt}\right) = k\varepsilon + c\frac{d\varepsilon}{dt}$$

where $k$ is the equivalent 'spring' stiffness and $c$ (=$\lambda k$) is the 'dashpot' constant.

The constant $\lambda$ is called the *time constant*; referring to the constitutive equation above when the material is strained and held at that strain $\varepsilon_0$ for a settling time and then released, then $\left(\varepsilon + \lambda^{d\varepsilon}/_{dt}\right) = 0$ or $\varepsilon(t) = \varepsilon_0\ \exp(-\lambda t)$. The time constant $\lambda$ thus determines the rate at which the material responds to load signals.

For cyclic straining, apply a periodic strain $\varepsilon(t) = \varepsilon_M + \varepsilon_A\ \sin\omega t$ where $\varepsilon_M$ is the mean strain, $\varepsilon_A$ is the strain amplitude, $t$ is time and $T$ is the circular frequency of the oscillation, radians per second, Fig. 2.11a.

The frequency $f = \omega/2\pi = 1/T$, where $T$ is the period. The corresponding load at the mean strain is $F_M = \left(F_f\varepsilon_M\right)/\varepsilon_f$ and the load increment corresponding to the strain amplitude $F_M = \left(F_f\varepsilon_M\right)/\varepsilon_f$.

This results in a transient fibre load

$$F(t) = \frac{F_f}{\varepsilon_f}\{\varepsilon_M + \varepsilon_A(sin\,\omega t + \lambda\,\omega cos\,\omega t)\}$$

or

$$F(t) = F_M + F_A\sqrt{1 + (\lambda\omega)^2}\,\sin(\omega t + \delta)$$

where the phase shift $\delta$ is given $\tan\delta = \lambda\omega$.

Thus the effect of viscoelasticity is firstly to amplify the load by the factor $\sqrt{1 + (\omega\lambda)^2}$. The maximum opening of the loop is $F_A\omega\lambda$, above and below the mean load. Secondly the viscoelastic parameter *8* causes to load to lead the strain, or alternatively the strain to lag behind the load by the phase shift $\delta$ in periodic motion by the phase.

**(a)**

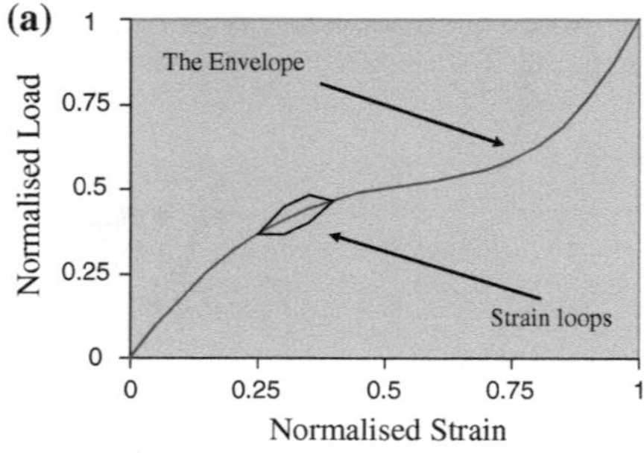


**(b)**

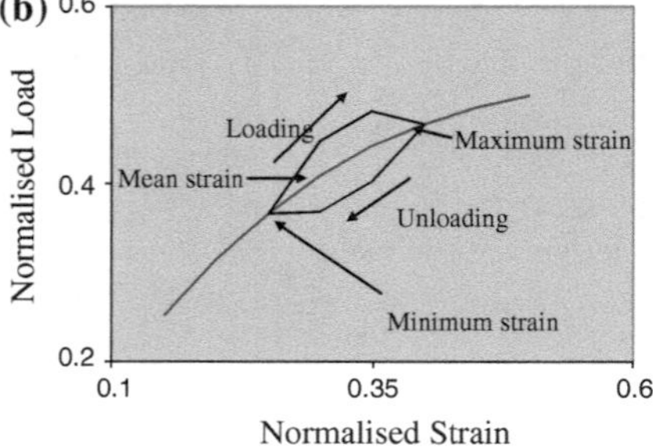


**Fig. 2.11** **a** Strain cycling. **b** Strain loops

Finally the presence of the loop, where the strain path is clockwise around the loop, lying above the static envelope for increasing strain and below for decreasing strain results in work done over the cycle, Fig. 2.11b.

The work done in moving from time (1) to time (2)

$$W = \int k\varepsilon d\varepsilon + \int c\frac{d\varepsilon}{dt}d\varepsilon = \left|k\frac{\varepsilon^2}{2}\right|_{\varepsilon_1}^{\varepsilon_2} + \int_{t_1}^{t_2} c\frac{d\varepsilon^2}{dt}dt$$
$$= W_E(\varepsilon_2, \varepsilon_1) + W_{VE}(t_2, t_1)$$

where $W_E$ is the elastic work done and $W_{VE}$ is the work done by the viscoelastic or rate terms.

Now using the expressions for strain $\varepsilon = \varepsilon_M + \varepsilon_A \sin(\omega t)$, and $d\varepsilon/dt = \omega\varepsilon_A \cos(\omega t)$, substituting in the work expressions above, applying to a path starting from the minimum point in the cycle, $\varepsilon = \varepsilon_M - \varepsilon_A$ and finishing at the maximum point, $\varepsilon = \varepsilon_M + \varepsilon_A$ (a semi cycle) results in the following

$$W = k\left[\frac{(\varepsilon_M + \varepsilon_A)^2 - (\varepsilon_M - \varepsilon_A)^2}{2}\right] + c\varepsilon_A^2\omega^2 \int_t^{t+T/2} \cos^2(\omega t)dt$$
$$= k[2\varepsilon_M\varepsilon_A] + c\,\varepsilon_A^2\omega^2\frac{T}{4}$$
$$= W_E(\varepsilon_M + \varepsilon_A, \varepsilon_M - \varepsilon_A) + \frac{W_{VE}\left(t + \frac{T}{2}, t\right)}{2}.$$

The work done by on the elastic component, $W_E$ is the area under the load extension curve, defined by the envelope, the strain axis, the lower limit, $\varepsilon = \varepsilon_M - \varepsilon_A$ and the upper limit, $\varepsilon = \varepsilon_M + \varepsilon_A$. $W_{VE}$ is the work done by the viscoelastic or rate terms over the whole cycle.

The work done on the viscoelastic components in a loop is

$$W_{VE} = c\omega^2\frac{T}{2}\varepsilon_A^2 = c\omega\pi\varepsilon_A^2 = k\,\tan\delta\,\pi\,\varepsilon_A^2$$

.

The ratio $\Re$ of these two work terms is

$$\Re = \frac{W_{VE}}{W_E} = \tan\delta\frac{\pi}{2}\frac{\varepsilon_A}{\varepsilon_M} = \omega\lambda\frac{\pi}{2}\frac{\varepsilon_A}{\varepsilon_M}$$

.

In summary, for linear viscoelasticity, the three properties, the load amplification, the work done within the cycle, and the phase shift between applied load and following strain * are interrelated and are all independent of the mean strain $\varepsilon_M$. For nonlinear viscoelasticity this is not so as the shift *, the cycle energy $W_{NE}$ and the load amplification $\sqrt{1 + (\omega\lambda)^2}$are all dependent on the strain reference $\varepsilon_M$.

### 2.5.3 Nonlinear Viscoelasticity

Recall that for a **linear** viscoelastic system, spring (stiffness or modulus $k$) and dashpot, $c$ in parallel with the spring, the load $L$ is given as

$$F = k\varepsilon + c\frac{d\varepsilon}{dt} = k\left(\varepsilon + \lambda\frac{d\varepsilon}{dt}\right)$$

where $\varepsilon$ is the strain and $t$ is time. For nonlinear viscoelasticity the load depends on a nonlinear combination of the strain and strain rate, that is $F = \Im\left(\varepsilon,\ {}^{d\varepsilon}/_{dt}\right)$. However as shown previously for small strain rates the first two terms of the Taylor series gives

$$F\left(\varepsilon, \frac{d\varepsilon}{dt}\right) = \Im(\varepsilon, 0) + \frac{d\varepsilon}{dt}\left|\frac{\partial\Im\left(\varepsilon, \frac{d\varepsilon}{dt}\right)}{\partial\left(\frac{d\varepsilon}{dt}\right)}\right|_{\frac{d\varepsilon}{dt}=0}$$

or quite generally

$$F\left(\varepsilon, \frac{d\varepsilon}{dt}\right) = F(\varepsilon) + G(\varepsilon)\frac{d\varepsilon}{dt}.$$

This is a quasilinear model, since this is separable and linear in strain rate but nonlinear in strain. Further development depends on the specification of the functions $F(\varepsilon)$ and $G(\varepsilon)$.

Here the form of the function $F(\varepsilon)$ is assumed to be polynomial in strain, that is

$$F(\varepsilon) = F_f\sum_{i=1}^{i=n} a_i\left(\frac{\varepsilon}{\varepsilon_f}\right)^i = F_f\sum_{i=1}^{i=n} a_i\,(\eta)^i \text{ and where } \sum_{i=1}^{i=n} a_i = 1$$

.

with the reversion to normalised strain, $\eta(=\varepsilon/\varepsilon_f)$.

To further develop the bridge from linear viscoelasticity, first recall the linear model

$$F\left(\varepsilon, \frac{d\varepsilon}{dt}\right) = k\varepsilon + c\frac{d\varepsilon}{dt} = k\left(\varepsilon + \lambda\frac{d\varepsilon}{dt}\right)$$

and the nonlinear variant

$$F\left(\varepsilon, \frac{d\varepsilon}{dt}\right) = F(\varepsilon) + G(\varepsilon)\frac{d\varepsilon}{dt}$$

.

A specific, but quite general form of the viscoelastic coefficient, the time constant $\delta(\varepsilon)$ can be introduced,

$$\lambda(\varepsilon)\frac{dF(\varepsilon)}{d\varepsilon} \equiv \lambda(\varepsilon)k(\varepsilon) = G(\varepsilon)$$

so that the constitutive model can now be written,

$$F\left(\varepsilon, \frac{d\varepsilon}{dt}\right) = F(\varepsilon) + \lambda(\varepsilon)\frac{dF(\varepsilon)}{d\varepsilon}\frac{d\varepsilon}{dt} = H(\varepsilon)\left(\varepsilon + \frac{\lambda(\varepsilon)}{H(\varepsilon)}\frac{dF(\varepsilon)}{d\varepsilon}\frac{d\varepsilon}{dt}\right)$$

where $H(\varepsilon)$, $(=F(\varepsilon)/\varepsilon)$ is the secant modulus. This is combined with the tangent modulus, $dF/d\varepsilon$ to define the modulus ratio $M_R$, as follows

$$M_R(\varepsilon) = \frac{1}{H(\varepsilon)}\frac{dF(\varepsilon)}{d\varepsilon} = \frac{dF(\varepsilon)}{d\varepsilon}\frac{\varepsilon}{F(\varepsilon)}.$$

The modulus ratio can be used as a measure of the nonlinearity of the elastic components; it is unity for linear elastic components for all values of strain. This definition is quite general since the *time 'constant'*, $\lambda(\varepsilon)$ has been employed to preserve the generality of the viscoelasticity component $G(\varepsilon)$.

The secant modulus $H(\varepsilon)$ can also assumed to be a polynomial. In order to develop the nonlinear viscoelastic model and to use the results from linear viscoelastic theory, a modulus function, the secant modulus $H(\varepsilon)$ introduced above is

$$H(\varepsilon) = \frac{F(\varepsilon)}{\varepsilon} = \left(\frac{F_f}{\varepsilon_f}\right)\sum_{i=1}^{i=n} a_i\left(\frac{\varepsilon}{\varepsilon_f}\right)^{i-1} = \left(\frac{F_f}{\varepsilon_f}\right)\sum_{i=1}^{i=n} a_i\,(\eta)^{i-1}$$

and recalling the tangent modulus,

$$\frac{dF(\varepsilon)}{d\varepsilon} = \left(\frac{F_f}{\varepsilon_f}\right)\sum_{i=1}^{i=n} a_i\,i\left(\frac{\varepsilon}{\varepsilon_f}\right)^{i-1} = \left(\frac{F_f}{\varepsilon_f}\right)\sum_{i=1}^{i=n} a_i\,i\,(\eta)^{i}$$

gives the modulus ratio

$$M_R(\varepsilon) = \frac{\sum_{i=1}^{i=n} a_i\,i\left(\frac{\varepsilon}{\varepsilon_f}\right)^{i-1}}{\sum_{i=1}^{i=n} a_i\left(\frac{\varepsilon}{\varepsilon_f}\right)^{i-1}} = \frac{\sum_{i=1}^{i=n} a_i\,i\,(\eta)^{i-1}}{\sum_{i=1}^{i=n} a_i\,(\eta)^{i-1}}.$$

Note that the secant modulus not singular, even though it includes $\varepsilon$ as a denominator, since it is assumed that there is a nonzero modulus at zero strain and that there is a stress-free, strain-free state, thus implying a leading $\varepsilon$ in the numerator (Fig. 2.12).

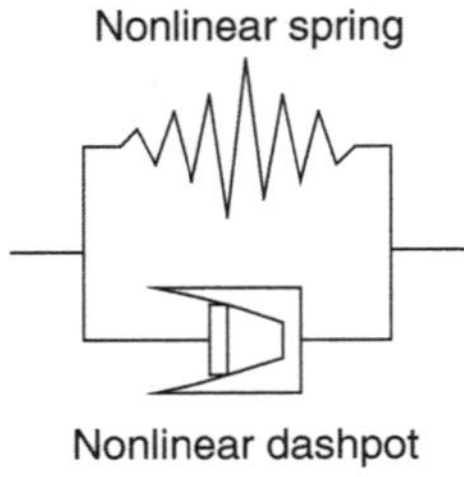


**Fig. 2.12** The nonlinear Voigt viscoelastic model

Introducing this function into the viscoelastic model,

$$F\left(\varepsilon, \frac{d\varepsilon}{dt}\right) = H(\varepsilon)\left\{\varepsilon + \lambda(\varepsilon)M_R(\varepsilon)\frac{d\varepsilon}{dt}\right\}.$$

Define now the *nonlinear time constant* $\Lambda(\varepsilon)$ as follows

$$\Lambda(\varepsilon_M) = \lambda(\varepsilon_M)M_R(\varepsilon_M)$$

then force function can be written

$$F\left(\varepsilon, \frac{d\varepsilon}{dt}\right) \approx H(\varepsilon)\left\{\varepsilon + \Lambda(\varepsilon)\frac{d\varepsilon}{dt}\right\}$$

For small oscillations, where the strain is $\varepsilon(t) = \varepsilon_M + \varepsilon_A \sin \omega t$ and where the cycle amplitude is small, $\varepsilon_A \ll \varepsilon_M$, the *nonlinear time constant* $\Lambda(\varepsilon)$ can be taken as constant over the cycle. Thus, the linearised form is

$$F\left(\varepsilon, \frac{d\varepsilon}{dt}\right) \approx H(\varepsilon_M)\left\{\varepsilon + \Lambda(\varepsilon_M)\frac{d\varepsilon}{dt}\right\}.$$

The phase shift, averaged over the cycle is now seen to be dependent on the mean strain $\varepsilon_M$, that is $\tan\delta \approx \Lambda(\varepsilon_M)\omega$.

For larger cycle amplitudes the dependency of $\Lambda$ on the strain $\varepsilon$ must be considered,

$$F\left(\varepsilon, \frac{d\varepsilon}{dt}\right)(\varepsilon_M + \varepsilon_A \sin \omega t)\{\varepsilon_M + \varepsilon_A \sin \omega t + \Lambda(\varepsilon_M + \varepsilon_A \sin \omega t)\omega\varepsilon_A \cos \omega t\}$$

and this is quite specific to the form of the functionals $H$ and $\Lambda$.

Similarly, the other two cycle parameters, the loop amplitude is $L_A T \Lambda$ above and below the mean load $L_M$, and the loop area, the work done on the viscoelastic components in a loop

$$W_{VE} = G(\varepsilon_A)\omega^2 \frac{T}{2}\varepsilon_A^2 = G(\varepsilon_A)\omega\pi\varepsilon_A^2 = H(\varepsilon_A)\tan\delta\,\pi\,\varepsilon_A^2.$$

Finally the ratio $\Re$ of these two work terms is

$$\Re = \frac{W_{VE}}{W_E} = \omega\frac{G(\varepsilon_A)}{H(\varepsilon_A)} = \tan\delta\frac{\pi}{2}\frac{\varepsilon_A}{\varepsilon_M} = \omega\Lambda(\varepsilon_A)\frac{\pi}{2}\frac{\varepsilon_A}{\varepsilon_M}$$

### 2.5.4 Large Strain Rate Modelling for Polymer Fibres

For high strain rate modelling the Taylor series approximation

$$F\left(\varepsilon, \frac{d\varepsilon}{dt}\right) = \Im(\varepsilon, 0) + \frac{d\varepsilon}{dt}\left|\frac{\partial\Im\left(\varepsilon, \frac{d\varepsilon}{dt}\right)}{\partial\left(\frac{d\varepsilon}{dt}\right)}\right|_{\frac{d\varepsilon}{dt}=0}$$

is not necessarily applicable; this approximation depends on the smallness of $d\varepsilon/dt$, and that the higher powers of strain rate $(d\varepsilon/dt)^2$, $(d\varepsilon/dt)^3$ etc. are negligible, so that the Taylor series can be truncated with the linear term. For these higher rates of strain, empirical fits are established asymptote to the static value for zero strain rates. The rate of loading affects the break point. Usually an extension test done at fast strain rate and results in a larger break load and a smaller break strain. Generally the load at a high strain rate is larger than that the load measured statically at that strain, Fig. 2.13.

It is not possible to establish the effects of high negative strain rates from a positive strain state nor to quantify their effect on the dynamic load.

Using the polynomial representation previously developed for fibre load,

$$F(\varepsilon) = F_f\sum_{i=1}^{i=n} a_i\left(\frac{\varepsilon}{\varepsilon_f}\right)^i = F_f\sum_{i=1}^{i=n} a_i\,(\eta)^i = F_f f(\eta).$$

and where the normalised strain, $\eta = \varepsilon/\varepsilon_f$. The load is thus dependent on the fracture load $L_f$ and the fracture strain $\varepsilon_f$

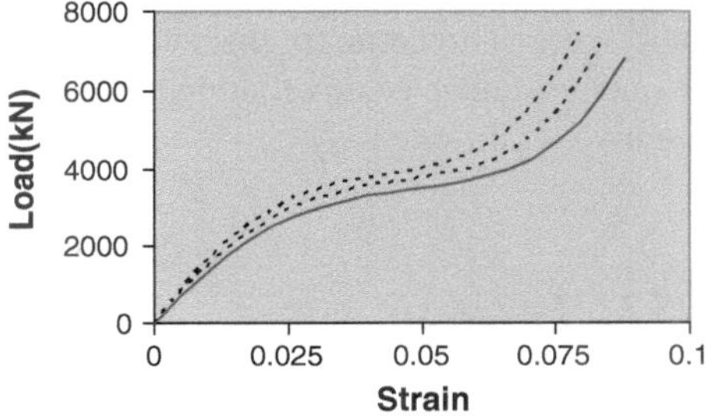


**Fig. 2.13** The effect of strain rate

To introduce the strain rate effect $\dot{\varepsilon}\left(= {}^{d\varepsilon}/_{dt}\right)$, let the two fracture parameters $F_f$ and $\varepsilon_f$ be functions of the strain rate, as following

$$F = F_f(\dot{\varepsilon})f(\eta) \quad \text{and} \quad \eta = \frac{\varepsilon}{\varepsilon_f(\dot{\varepsilon})}.$$

It now remains to specify the functions $L_f$ and $\varepsilon_f$; these functions must determined by fitting to various experimental data. Two proposed but not exclusive empirical fits follow, first

$$F_f(\dot{\varepsilon}) = F_{ref}\left[1 + \lambda\sqrt{ln\left(\frac{\dot{\varepsilon}}{\dot{\varepsilon}_{ref}}\right)}\right]$$

$$\varepsilon_f(\dot{\varepsilon}) = \varepsilon_{ref}\left[1 + \gamma\sqrt{ln\left(\frac{\dot{\varepsilon}}{\dot{\varepsilon}_{ref}}\right)}\right]$$

where $F_{ref}$ and $\varepsilon_{ref}$ are the breaking load and breaking strain at a reference strain rate $\dot{\varepsilon}_{ref}$. The constants $\lambda(\approx 0.2, 0.05)$ for nylon and polyester respectively and $\gamma$ ($\approx$ –0.03, 0.01) are empirically derived constants. This fit is applicable for nylon (0.000013 $s^{-1} < \dot{\varepsilon} <$ 10.96 $s^{-1}$) and for polyester (0.000166 $s^{-1} < \dot{\varepsilon} <$ 1.66 $s^{-1}$).

Secondly, for an aramid, Twaron the following fit has been established,

$$F_f = F_{ref}\frac{1 + \alpha\exp\left(\lambda\frac{\dot{\varepsilon}}{\dot{\varepsilon}_{ref}}\right)}{1 + \alpha\exp(\lambda)}$$

$$\varepsilon_f = \varepsilon_{ref}\frac{1 - \beta\exp\left(\lambda\frac{\dot{\varepsilon}}{\dot{\varepsilon}_{ref}}\right)}{1 - \beta\exp(\gamma)}$$

for large strain rates, in the range 470 $s^{-1} < \dot{\varepsilon} <$ 1100 $s^{-1}$, where the reference strain rate, $\dot{\varepsilon}_{ref}$ (=0.0167 $s^{-1}$). $\alpha, \beta, \lambda$ and $\gamma$ are empirically derived material constants, typically 0.04, 0.02, $7 \times 10^{-5}$ and $4.8 \times 10^{-5}$ respectively.

The empirical fits shown above are not unique; there are many other fits that could be used.

Although the second empirical formula could be applied for negative strain rates, there is no certainty that such would be useful. Recall that since the tests are to fracture under positive strain rate; it is not possible to determine the effect of negative strain rate.

### 2.5.5 *Hysteresis*

Some polymers show an effect similar to that for metals when loaded past the yield point. To illustrate this effect, the static load extension curve $F = F_f f(\eta)$ is defined as the envelope, where $F$ is the fibre load, and $\eta = \varepsilon/\varepsilon_f$ is the normalised strain, bounded $0 \leq \eta \leq 1$. The fibre fracture load is $F_f$, and $\varepsilon_f$ is fibre maximum, fracture or break strain. The normalised modulus is defined by the following,

$$\frac{df}{d\eta} = \frac{dF}{d\varepsilon}\frac{\varepsilon_f}{F_f}$$

Consider the loading sequence where the fibre is subject to an increasing strain to a final specific strain. This is defined as the set strain, $\varepsilon_s$, Fig. 2.14 and the corresponding normalised set strain is $\eta_s$. At this stage if the strain is then reduced and the load does not recover along the envelope, but some other path below the envelope the system is memory dependent or hysteretic, it remembers the set strain and this now governs the new unloading and subsequent loading paths. The equivalent in metals is the yield point and plastic unloading. Thus the material is also anelastic, violating the one to one correspondence between load and strain. The unloading path, the spline lies below the envelope until a point is reached when either the unloading is stopped or when the load is zero. At this latter point the strain is defined as the residual strain, $\eta_r$.

If the material is cycled between two strains, the larger being the set strain, so that the cycling load extension curve is along the spline, the apparent modulus is the slope of the spline which is larger than that along the envelope, Fig. 2.15. This increase in modulus is called modulus modulation, $M$.

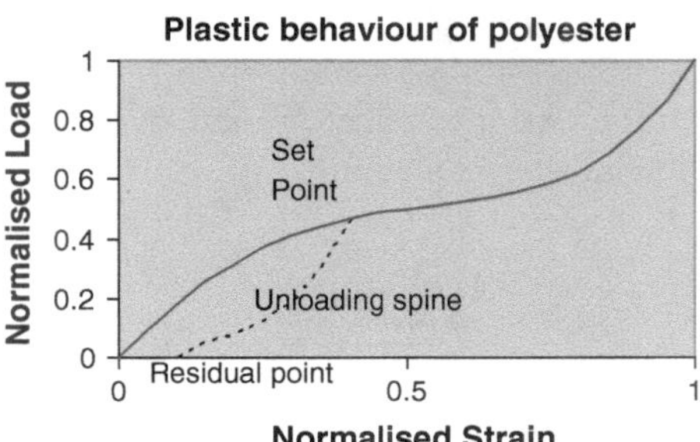


**Fig. 2.14** The envelope and unloading spline

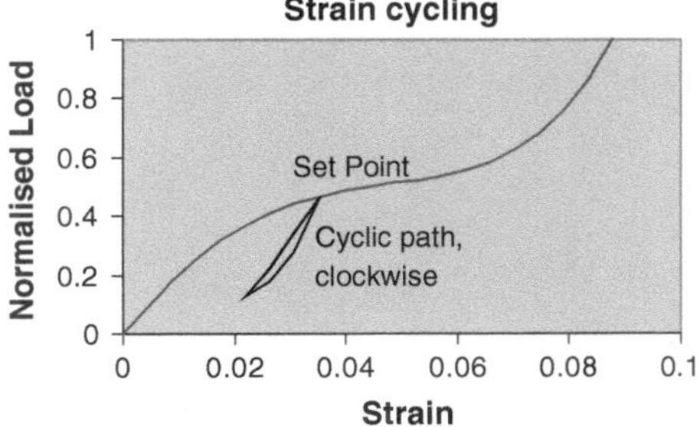


**Fig. 2.15** Cycling along the spline, showing modulus modulation

The normalised modulus, $df/d\eta$ is shown in the following figure; also shown is a typical modulated modulus. This is the increased modulus resulting from cyclic loading on the unloading spine for different set strain. The modulation varies with the set strain, and a candidate polynomial fit can be applied to model of the modulation,

$$M = a_0 + a_1\eta_s + a_2\eta_s^2$$

Typical values of these coefficients, $a_0$, $a_1$ and $a_2$ are 1, 7 and −5 respectively. The unloading path from the set strain is also approximated by a polynomial,

$$\frac{F}{F_f} = (\varepsilon - \varepsilon_r)(A + B(\varepsilon - \varepsilon_r))$$

or in terms of the normalised strain

$$f_s(\eta) = (\eta - \eta_r)(A + B(\eta - \eta_r))$$

where $A$ and $B$ are local constants which depend on the set point and the modulus modulation, and $\varepsilon_r$ is the residual strain, the strain left when the yarn in unloaded (Figs. 2.16, 2.17 and 2.18).

The normalised residual strain $\eta_r$ ($=\varepsilon_r/\varepsilon_s$) is also a function of the normalised set strain $\eta_s$ and again is approximated by a candidate polynomial fit

$$\eta_r = b_1\eta_s + b_2\eta_s^2$$

where again typical values for the coefficients $b_1$, and $b_2$ are typically 0.7 and −0.2.

The two coefficients $A$ and $B$ are extracted as solution to the following pair simultaneous equations

$$\begin{aligned} f_s(\eta_s) &= (\eta_s - \eta_r)(A + B(\eta_s - \eta_r)) = f(\eta_s) \text{ and} \\ \frac{df_s(\eta_s)}{d\eta} &= A + 2B(\eta_s - \eta_r) = M(\eta_s)\frac{df(\eta_s)}{d\eta} \end{aligned}$$

so that

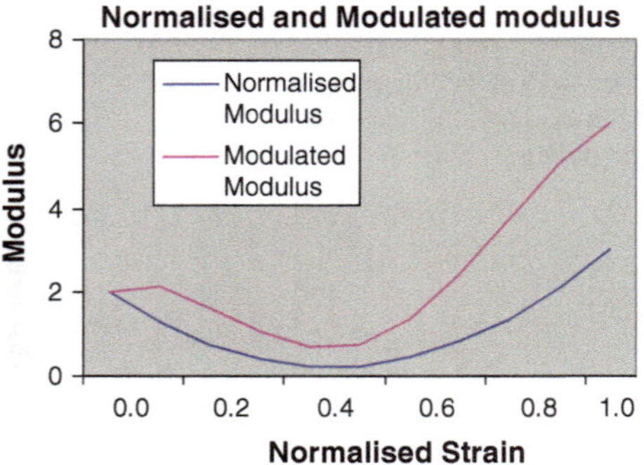


**Fig. 2.16** Modulus modulation

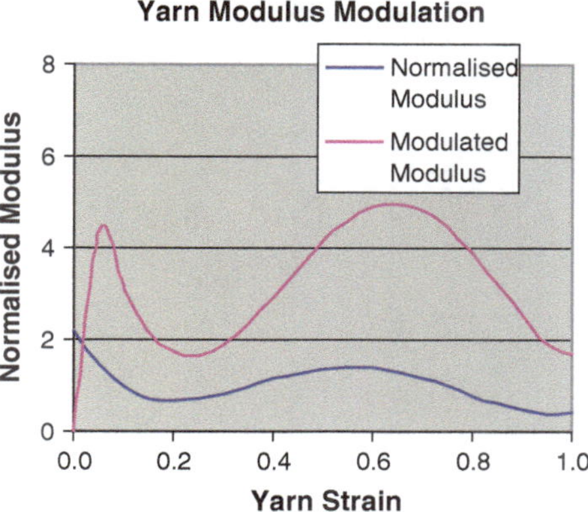


**Fig. 2.17** Normalised and modulated modulus

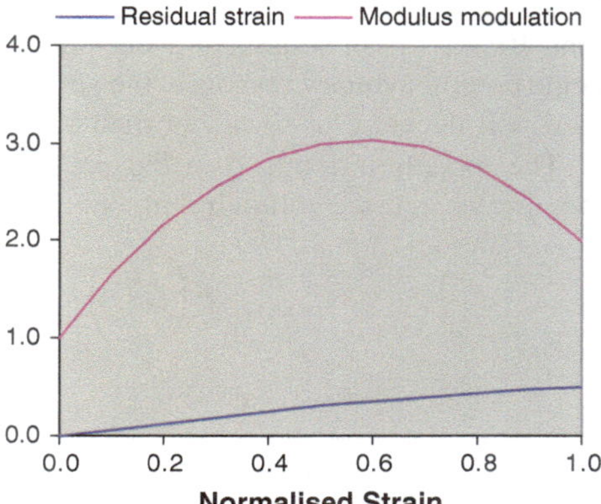


**Fig. 2.18** Residual strain

$$A = \frac{2f(\eta_s)}{(\eta_s - \eta_r)} - M(\eta_s)\frac{df(\eta_s)}{d\eta}$$
$$B = \frac{1}{(\eta_s - \eta_r)^2}\left\{(\eta_s - \eta_r)M(\eta_s)\frac{df(\eta_s)}{d\eta} - f_s(\eta_s)\right\}.$$

Figure 2.19 shows the normalised load plotted against the normalised strain, and also the unloading, recovery path for a set strain 0.4.

## 2.5.6 Creep and Relaxation

The previous section has been focussed on the response of fibre materials to strain rates, from slow rates, 0.01 $s^{-1}$ to high rates, 1,000 $s^{-1}$. However there is another domain, very small strain rates which needs to be addressed, and these are

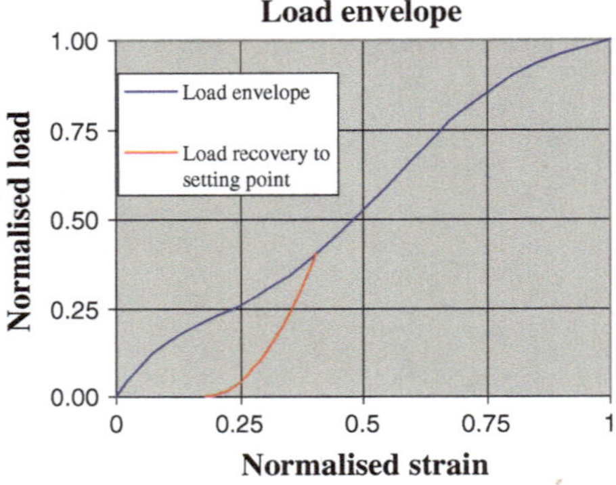


**Fig. 2.19** Unloading spine, normalised set strain = 0.4

classified by creep and relaxation. Creep is the strain response to a constant sustained load over a long period, and results in strain rates typically substantially less than $10^{-5}\ s^{-1}$. For some load values the materials will creep to failure. Relaxation is the complementary response, the strain is sustained for an extended time and the load will decrease to some constant value (Figs. 2.20 and 2.21).

The models introduced in the previous sections are responsive to the higher strain rates and the following model

$$F\left(\varepsilon, \frac{d\varepsilon}{dt}\right) = F(\varepsilon) + G(\varepsilon)\frac{d\varepsilon}{dt}$$

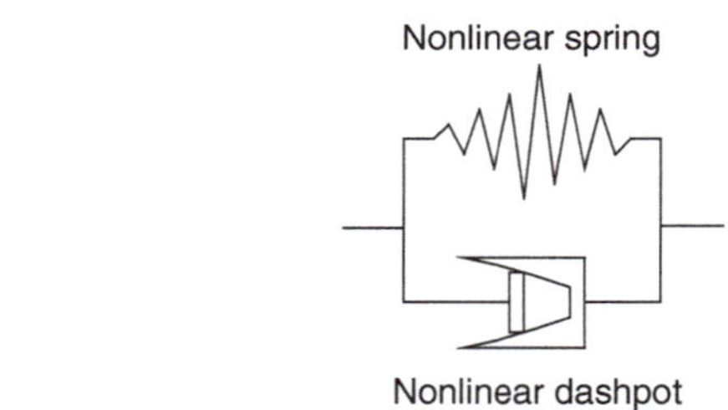


**Fig. 2.20** The nonlinear viscoelastic model

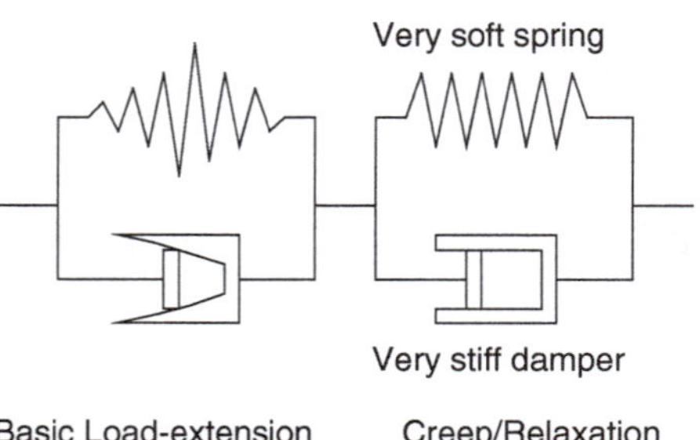


**Fig. 2.21** The creep-relaxation viscoelastic model

can exhibit creep behaviour; for a given load $F_{constant}$, the material will creep until the strain rate $d\varepsilon/dt$ is zero and the limit strain $\varepsilon_{limit}$ is reached

$$F_{constant} = F(\varepsilon) + G(\varepsilon)\frac{d\varepsilon}{dt} = F(\varepsilon_{limit}).$$

This model can be solved for time as a function of strain

$$\int_{\varepsilon=0}^{\varepsilon} \frac{G(\varepsilon)d\varepsilon}{F_{constant} - F(\varepsilon)} = t(\varepsilon)$$

This function could be inverted, as least numerically to give $\varepsilon(t)$. For a given load $F_{constant}$, the material will creep until the strain rate $d\varepsilon/dt$ is zero and the limit strain $\varepsilon_{limit}$ is reached

$$F_{constant} = F(\varepsilon) + G(\varepsilon)\frac{d\varepsilon}{dt} = F(\varepsilon_{limit}).$$

This model will not yield relaxation as the application of the load $F_{constant}$, is fairly immediate and the load is solely carried by the functional $F(\varepsilon)$. To develop the model so that it will encompass both the short time, shock and cyclic loading and the long time, creep and relaxation behaviours, a secondary, slow viscoelastic model is introduced in series with the primary, fast model,

$$F\left(\varepsilon, \frac{d\varepsilon}{dt}\right) = F_p(\varepsilon_p) + G_p(\varepsilon_p)\frac{d\varepsilon_p}{dt} = F_s(\varepsilon_s) + G_s(\varepsilon_s)\frac{d\varepsilon_s}{dt}$$

where the strain $\varepsilon$ is now the sum of the two constituent strains, the primary strain $\varepsilon_p$ which is the response to fast loading profiles and the secondary strain $\varepsilon_s$ which is the contribution to strain from the slow or sustained loading.

The material strain and strain rate are thus

$$\varepsilon = \varepsilon_p + \varepsilon_s \quad \text{and} \quad \frac{d\varepsilon}{dt} = \frac{d\varepsilon_p}{dt} + \frac{d\varepsilon_s}{dt}.$$

To tune the various components so that the two regimes can be included, the various constituents are such that, for a given strain

$$F_p(\varepsilon) < < F_s(\varepsilon) \quad \text{and} \quad G_p(\varepsilon) > > G_s(\varepsilon).$$

Consider an applied force, applied to the fibre material

$$F_{applied}(t) = F_p(\varepsilon_p) + G_p(\varepsilon_p)\frac{d\varepsilon_p}{dt} = F_s(\varepsilon_s) + G_s(\varepsilon_s)\frac{d\varepsilon_s}{dt}.$$

Thus since the 'spring' in the secondary component is much larger that the 'spring' component in the primary component, the majority of strain in the material is primary, $\varepsilon \approx \varepsilon_p$. The 'visco' term in the secondary component is much smaller than the primary 'visco' component so that the secondary component contributes little to the fast time loading. However in the slow time modelling, the primary model reaches equilibrium quickly

$$F_{applied}(t) = F_p(\varepsilon_p) + G_p(\varepsilon_p)\frac{d\varepsilon_p}{dt}$$

and the secondary model will continue straining, again according to the following equation

$$F_{applied}(t) = F_s(\varepsilon_s) + G_s(\varepsilon_s)\frac{d\varepsilon_s}{dt}.$$

The functionals $F_p(\varepsilon)$, $F_s(\varepsilon)$, $G_p(\varepsilon)$ and $G_s(\varepsilon)$ must be established experimentally. The primary functionals are determined using static loading and cyclic testing whereas the secondary functionals are determined from creep and relaxation testing.

In the creep phase, the creep or secondary strain is much larger that the primary strain. In relaxation where the total strain is fixed there is a continued readjustment between the primary strain and secondary strain. Initially the secondary strain is zero and the total strain is the primary strain; during relaxation the primary strain decreases and the secondary strain increases. However since $F_p(\varepsilon) << F_s(\varepsilon)$, the load will eventually be the secondary load.

Creep response to a constant load is often characterised by three zones, primary, secondary and final phases as zones against time. However since there is a correspondence between time and creep or secondary strain these zones can be correlated against creep strain.

It is often more useful to represent the secondary model by a spring dashpot model

$$\begin{aligned}F\left(\varepsilon, \frac{d\varepsilon}{dt}\right) &= F_s(\varepsilon_s) + G_s(\varepsilon_s)\frac{d\varepsilon_s}{dt}\\ &= k(\varepsilon_s)\varepsilon_s + c(\varepsilon_s)\frac{d\varepsilon_s}{dt}\\ &= k(\varepsilon_s)\left(\varepsilon_s + \lambda(\varepsilon_s)\frac{d\varepsilon_s}{dt}\right)\end{aligned}$$

where both the coefficients $k(\varepsilon_s)$ and $c(\varepsilon_s)$ or $8(\varepsilon_s)$ are again measured against the secondary strain.

## 2.6 Secondary Deformations

The principal attribute of a fibre is its length and its main structural resistance is associated with the axial stress or in line tension; the secondary actions specified here are torsion/twist and moment/flexure reactions and deformations [4–6]. They are equally important within the general consideration of fibre performance although in specific geometries or applications one may be more important than the other. Torsion/twist behaviour is considered first since in general it is simpler.

## 2.6.1 Fibre Torsion

The theory outlined here is the classical torsion theory as applied to a circular shaft. If the fibre is considered cylindrical and torques are applied equally and opposite at each end of the fibre the fibre is in overall equilibrium; to resist the torsion the fibre must deform in a twist fashion, Fig. 2.22.

Consider a cylindrical shaft and specifically a shell at a constant radius from the fibre axis; now imagine the development formed by slitting along a generator on the surface and parallel to the fibre axis. This development will deform in a *simple* shear mode when the fibre is twisted; in this mode the shear results in one set of edges, the generators stretching, although for a small twist this stretch is assumed negligible. In comparison *pure* shear is accompanied by equal stretching of both sets of generators.

Considering the simple shear of the cylindrical shell and small twists, the shear strain $\gamma$ is approximated by $r\theta/L$ where $r$ is the radial station of the shell, $L$ is the axial length, the thickness of the shell is $dr$ and $\theta$ is the relative rotation of the fibre sections defined by the two edges of the shell. The shear stress $\tau$ developed in the shell is $G\gamma$ where $G$ is the shear modulus and the torque developed due to this shell is $2\pi r\tau dr$. To further develop the theory the following are assumed:

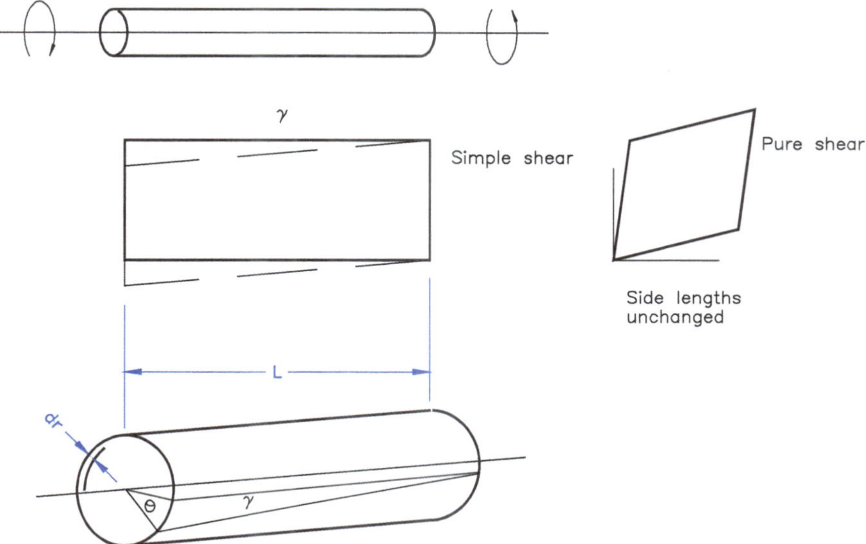


**Fig. 2.22** Fibre torsion and the development into simple shear

i. the plane cross sections rotate as rigid planes and
ii. there is no axial displacement.

The two assumptions ensure that any radial generator remains straight, perpendicular to the axis and does not move along the fibre axis. This removes from the possible deformation any warping of the cross section; whereas warping is inevitable, because of the dimensional characteristic of a fibre where $d \ll L$, this assumption is reasonable. However for non-circular sections warping is present but the resulting equations will be similar to those for circular sections.

The torsion is determined by considering the cylindrical shell, above. The shear stress in that shell ($\tau$) times the moment ($r$) times area ($2\pi r dr$) contributes to a reaction torque $dT$. The total torque is then

$$T = \int_0^{d/2} 2\pi \tau r^2 dr$$

where $d$ is the diameter of the fibre; substituting for the shear stress above gives

$$T = \frac{GJ\theta}{L}$$

where $J$ is the second moment of area of the cross section about the axis of rotation (polar moment of area) and is given as $J = \frac{\pi d^4}{32}$. The ratio $\theta/L$ is the twist (radians/unit length) of the fibre.

## 2.6.2 Fibre Flexure

The flexure or bending of fibres is based upon the classical bending of beams; in the first instance there is infinity of axes about which the fibre can bend, Fig. 2.23 To identify the mode or axis of bending, it is noted that the cross section has various attributes; first there is a centre of area, a position in the cross section and there is an area quantity. Secondly there are second moments of area, about axes through the centre of area and in the plane of the section; there is an infinite number of these axes and it is usual to concentrate on two orthogonal axes. As well as the second moments of area, there is a product moment of area and this quantifies the coupling between moment about one axis and flex about an orthogonal axis. It is possible to eliminate this coupling by choosing a specific pair of orthogonal axes and these are called *principal axes*. In the development that follows it is initially assumed that flexure occurs about a principal axis and that the developed moment will also occur about the same axis. The fibre, initially straight is bent by the application of a pair of equal and opposite bending moments $M$ at the ends of a considered section.

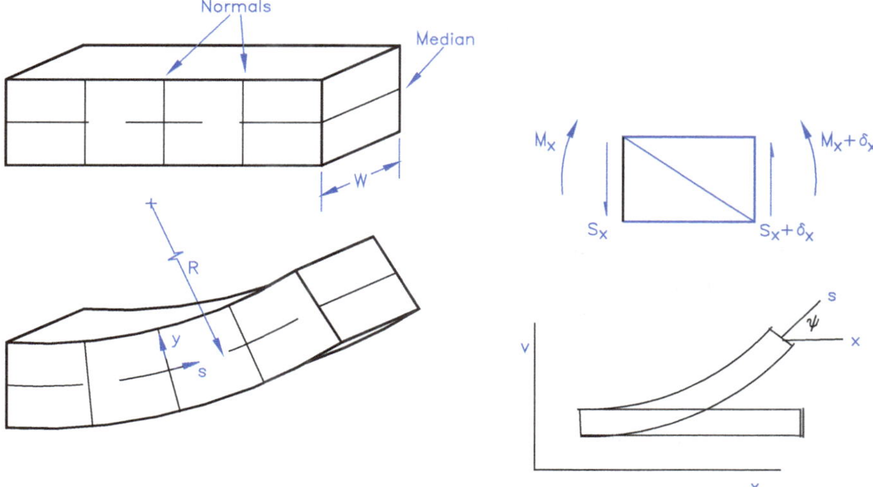


**Fig. 2.23** The flexure of a fibre, showing the normal and the median

The *median* or the axis along the fibre through the centre of area is bent and the angle made by the median and the initial fibre axis is $\psi$; $s$ is a coordinate running along the fibre axis and the curvature $\chi$ of the fibre will be given by $\chi = \frac{d\psi}{ds}$.

There is a differential straining of element lines parallel to the median, those below the median being in tension and those above in compression; an anti-clockwise applied moment $M$ is opposed by the moment of the reactive stresses about the median

$$M = -\int_{l}^{u} w\sigma y dy.$$

The upper and the lower limits $u$ and $l$, denote the top and bottom surface of the fibre and $w(y)$ is the width of the section, as a function of the position y, above the median.

At a position $y$ above the median, the strain $\varepsilon$ is $-y\chi$, compressive. If a linear stress–strain constitutive relation is used, so that the material stress $\sigma$ is the associated $E\varepsilon$, and substituting for the stress in the above expression, the applied moment becomes

$$M = E\chi \int_{l}^{u} w y^2 \, dy$$

The second moment of area about the flex axis (orthogonal to both the $y$ and the fibre axis) is

$$I_{zz} = \int_{l}^{u} w y^2 \, dy$$

and the developed bending moment resulting from the imposition of flexure is $M_z = EI_{zz}\chi_z$. There is a similar expression for flexure about the $y$ axis, $M_y = EI_{yy}\chi_y$. In each the subscript $y$ or $z$ denotes the axis about which the bending moment and curvature is effected and the axis about which the second moment of area is evaluated.

The above assumes linear elasticity, that is that the stress, $\sigma$ is proportional to the strain, $\varepsilon$; if the fibre constituent material is nonlinear with $\sigma = \sigma_f F(\varepsilon)$ then the bending moment becomes $M = \sigma_f \int_l^u F\left(\frac{\chi y}{\varepsilon_f}\right) wydy$. If a polynomial fit is assumed, $\sigma = \sigma_f \sum_{i=1}^{i=n} a_i \eta^i$, then $M = \sigma_f \sum_{i=1}^{i=n} \frac{a_i \chi^i}{\varepsilon_f^i} \int_l^u w(y)\, y^{i+1}\, dy$. Now consider for this section a fibre bent only in the global plane $xy$; the angle $\psi$ denotes the slope of the median to the $x$ axis and $v$ is the vertical position of the fibre median. The variables $x(s)$ and the variable $v(s)$ are recovered by $\frac{dx}{ds} = \cos\psi$ and $\frac{dv}{ds} = \sin\psi$. The fibre path is then defined by the moment distribution $M(s)$; this system of equations is often solved in terms of $s$, the running coordinate.

It is not always convenient to use the running coordinate $s$ since the location of the fibre under load is not known 'a priori'; instead a global space coordinate, $x$ can be employed by converting $x(s)$ and $v(s)$ to $s(x)$ and $v(x)$. Recalling the definition of the curvature $\chi$, and differentiating the above expression for $dv/ds$ yields an alternate expression for $\chi$ as $\chi = \frac{1}{\cos\psi}\frac{d^2 v}{ds^2}$. But $\frac{d}{ds} = \frac{dx}{ds}\frac{d}{dx} = \cos\psi \frac{d}{dx}$ and after a little manipulation the curvature for large deformations is usually given as $\chi = \frac{\frac{d^2 v}{dx^2}}{\left[1+\frac{dv}{dx}^2\right]^{3/2}}$.

The angle $\psi$ has been used to represent the inclination of a planar deformation with a initial axis; if this angle is small such that $dx/ds \approx 1$ and $dv/ds \approx \psi$, then for linear elasticity, the curvature is replaced by the following approximation

$$\chi \approx \frac{d^2 v}{dx^2}.$$

This leads to the classical Euler–Bernoulli beam theory where, the moment is given by $M = EI_{zz}\chi$

$$\frac{d^2 v}{dx^2} = \frac{M}{E I_{zz}}.$$

## 2.7 Strain Energy or the Energy of Deformation

Previously the strain energy arising from the axial extension of a fibre has been the result of an applied stress $\sigma$. At that point simple fibres were defined as those without coupling between the deformation modes. Here this is developed further by considering the strain energy and as such the class of fibre materials must be elastic or more correctly *hyperelastic*; the distinction between these classes will not be pursued here and the attribute elastic will be used throughout. A complete set of definitions and the distinction between elastic and hyperelastic materials is found in modern elasticity theory; however, for completeness a material is said to be hyperelastic if it possesses a strain energy functional which is an analytic function of strain and has a stress-free strain-free state.

For a given deformation state of a fibre, there exists a function, $W_l\left(\varepsilon, \frac{d\theta}{ds}, \chi\right)$, the units being energy/unit length. This associates the strain energy with the deformation state, governed by the strain $\varepsilon$, the twist/unit length $d\theta/ds$, and the curvature $\chi(=d\psi/ds)$. The reactive or developed generalised forces to sustain the deformed state are

$$\sigma = \frac{1}{A}\frac{\partial W_l}{\partial \varepsilon}, \quad T = \frac{\partial W_l}{\partial\left(\frac{d\theta}{ds}\right)} \quad \text{and} \quad M = \frac{\partial W_l}{\partial \chi}$$

where $A$ is the fibre cross section area. The associated units for stress, $\sigma$ are load/area, (e.g. N/m$^2$) and for torque and moment they are Nm.

If instead the units of $W_d$ are energy/weight, that is J/kg then

$$\sigma = \frac{\partial W_d}{\partial \varepsilon}, \quad T = \frac{\partial W_d}{\partial (d\theta/ds)} \quad \text{and} \quad M = \frac{\partial W_d}{\partial \chi},$$

and the associated units for stress become load/weight per unit length, (e.g. N/tex, or specific stress) and for torque or moment become couple/weight per unit length, (e.g. Nm/tex or specific torque, moment).

The coupling of the deformation modes, that is the effect of stretch on torque and moment, the effect of twist on tension and moment and the effect of flexure on the tension and torque are determined from the following coupling partial differentials,

$$\frac{\partial^2 W_d}{\partial \varepsilon \partial (d\theta/ds)}, \frac{\partial^2 W_d}{\partial \varepsilon \partial \chi} \quad \text{and} \quad \frac{\partial^2 W_d}{\partial (d\theta/ds) \partial \chi}.$$

If all the above differentials are zero, there is no coupling and the fibre is *simple* and a simple fibre has the following form of energy functional,

$$W_d\left(\varepsilon, \frac{d\theta}{ds}, \chi\right) = W_{d,e}(\varepsilon) + W_{d,t}\left(\frac{d\theta}{ds}\right) + W_{d,b}(\chi),$$

where the subscripts *e*, *t* and *b* refer to the extensional, twist and flexure modes.

The source of coupling between the extension, twist and flexure coupling can be expanded by considering a single homogeneous fibre. If in the course of the fibre production there is a twisting process, there will be a twist or helical anisotropy inherent in the fibre material; alternatively a 'fibre' could be a yarn type of structure which is assembled from a group of filaments twisted together. In these fibre structures, a stretch will cause a tendency to twist and if this is resisted a torque will be developed; alternatively if the fibre structure is twisted either with or against the twist anisotropy, the structure will shorten or lengthen and again if this is resisted, there will be a developed tension or compression.

For a homogeneous fibre, whose material properties are uniform across the section, an extension applied to the fibre will produce a uniform strain and thus stress. There is thus no flexure produced by extension. Similarly a fibre material whose stress strain behaviour is odd in strain will not develop a tension due to flexure. A material is odd in strain if the stress resulting from a negative, compressive strain is the same but negative to that produced from the same but tensile strain. This is algebraically expressed $\sigma(\eta) = -\sigma(-\eta)$ and for a polynomial representation of stress, the form contains only the odd powers, $\sigma = \sigma_f \sum_{i=1}^{i=n} a_i \, \eta^{2i-1}$ .

However a nonlinear fibre material which has different stress strain behaviour in compression to tension, that it contains in the polynomial form some even powers of the strain, $\sigma = \sigma_f \sum_{i=1}^{i=n} a_i \, \eta^i$, will produce develop a reactive tension due to flexure.

## References

1. Hearle JWS (1982) Polymers and their properties: fundamentals of structure and mechanics, vol 1. Ellis Horwood, Chichester
2. Hearle JWS, Thwaites JJ, Amirbayat J (1980) Mechanics of flexible fibre assemblies, NATO Advanced Study Institute Series, series E: applied sciences-No.38. Sijthoff & Noordhoff, Germantown
3. Billington EW (1986) Introduction to the mechanics and physics of solids. Adam Hilger Ltd, Boston
4. Williams JG (1980) Stress analysis of polymers, 2nd(revised) edn. Ellis Horwood Series in Engineering Science, Chichester
5. Mase GT, Mase GE (1999) Continuum mechanics for engineers, 2nd edn. CRC Press Ltd, Boca Raton
6. D'Souza AF, Garg VK (1984) Advanced dynamics, modeling and analysis. Prentice-Hall, New Jersey

# Chapter 3
# Component Path Geometries

**Abstract** This chapter considers the general fibre paths and specifically helical geometry; the effect of deformation of the helix on the deformation of the constituent fibre component is analysed. Finally a comparison in the energy of extension, twist and flexure is made so that useful engineering approximations can be selected.

Prior to considering in detail the assembly of components within a structure it is essential to consider the mathematics of a component path [1], with this in mind the vectorial representation is introduced although it is the end result that is important. First a general curve is represented, then the helix is introduced and the concept of geometry preservation is discussed. The energy required to deform the helix is then quantified and partitioned into that associated with component stretch, twist and flexure; this is important since it justifies the assumptions made on the relative importance of the deformation modes.

## 3.1 Curves in Space

### *3.1.1 General Path Geometry*

Consider a fibre path in three dimensional space (x, y, z) and the vector triplet ($\underline{i}$, $\underline{j}$ and $\underline{k}$); then a radius vector $\underline{r}$ is as follows

$$\underline{r} = x(\xi)\underline{i} + y(\xi)\underline{j} + z(\xi)\underline{k}$$

where $\xi$ is a parameter that is connected to the position along the trajectory, Fig. 3.1.

The **unit tangent** vector $\underline{u}$ is given by

$$\underline{u} = \frac{d\underline{r}}{ds} = \left(\frac{dx}{d\xi}\underline{i} + \frac{dy}{d\xi}\underline{j} + \frac{dz}{d\xi}\underline{k}\right)\frac{d\xi}{ds},$$

C. M. Leech, *The Modelling and Analysis of the Mechanics of Ropes*,
Solid Mechanics and Its Applications 209, DOI: 10.1007/978-94-007-7841-2_3,

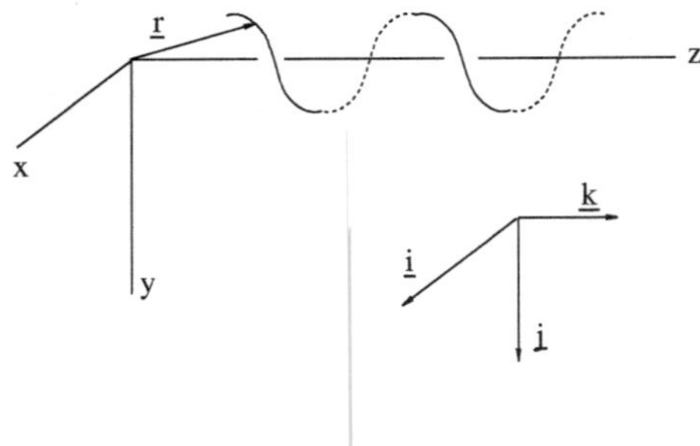

**Fig. 3.1** The general path of a curve in 3D space

which aligns itself locally with the direction of the fibre, Fig. 3.2. The curvature $\chi(s)$ of the path is $\chi(s) = \left|\frac{d\underline{u}}{ds}\right|$ and the **principal unit normal** (unit curvature vector) $\underline{p}$ is thus $\underline{p} = \frac{1}{\chi(s)}\frac{d\underline{u}}{ds}$,

The principal unit normal establishes the osculating **plane** as that which instantaneously contains both the principal normal $\underline{p}$ and the unit tangent $\underline{u}$ and is tangent to the curve at that location $\underline{r}$. The third vector, orthogonal with the unit tangent vector and the unit curvature vector is the **unit binormal** $\underline{b}$; this vector, the unit binormal is formed from a vector product of the other unit vectors as follows,

$$\underline{b} = \underline{u} \wedge \underline{p}$$

These three unit vectors form a moving trihedral or a vector basis that follows the fibre path. The osculating plane has been defined above as that plane that contains the principal normal $\underline{p}$ and the unit tangent $\underline{u}$ and is tangent to the fibre line at $\underline{r}$; other planes can be defined, namely the **tangent plane** that contains the unit tangent $\underline{u}$ and the binormal $\underline{b}$ and is also tangent to the curve at $\underline{r}$ and the **section plane** that contains the principal normal $\underline{p}$ and the binormal $\underline{b}$ and is normal to the curve at $\underline{r}$.

Now since $\underline{b} \cong \underline{u} = 0$ then $\frac{d\underline{b}}{ds} \cdot \underline{u} + \underline{b} \cdot \frac{d\underline{u}}{ds} = 0$. But $\underline{b} \cdot \frac{d\underline{u}}{ds} = 0$ since d$\underline{u}$/ds is in the direction of $\underline{p}$. Then $\frac{d\underline{b}}{ds} \cdot \underline{u} = 0$ or $\frac{d\underline{b}}{ds}$ is perpendicular to both $\underline{u}$ and to $\underline{b}$, so that $\frac{d\underline{b}}{ds} = -\tau\underline{p}$ where $\tau$ is a scalar, called the torsion or tortuosity of the curve at that point. The torsion is thus $\tau = -\underline{p} \cdot \frac{d\underline{b}}{ds}$.

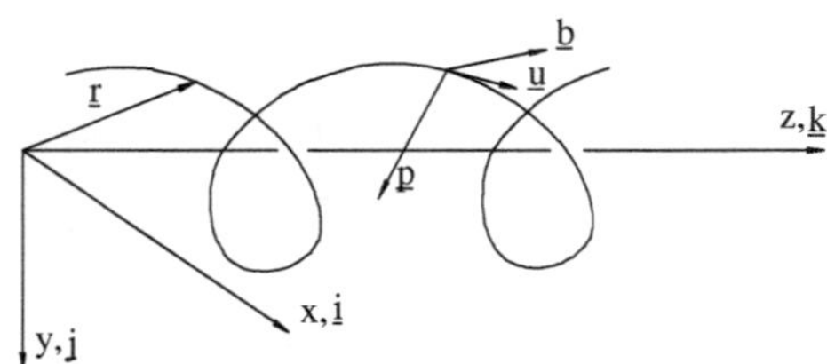

**Fig. 3.2** 3D curve, the units tangent, normal and binormal

The Frenet-Serret equations then follow

$$\frac{d\underline{u}}{ds} = \chi\underline{p}, \quad \frac{d\underline{p}}{ds} = -\chi\underline{u} + \tau\underline{b} \quad \text{and} \quad \frac{d\underline{b}}{ds} = -\tau\underline{p}.$$

since

$$\underline{p} = \underline{b} \wedge \underline{u}, \quad \underline{b} = \underline{u} \wedge \underline{p} \quad \text{and} \quad \underline{u} = \underline{p} \wedge \underline{b}$$

### *3.1.2 Geometry Preserving Structures*

The geometry of a fibre path is defined by equations and by parameters within the equations; if as a result of the structure deformation the path equations do not change but the parameters do, reflecting the amount of structure deformation then the deformation is said to be *geometry preserving*; this assumption ensures that the initial fibre crossover locations are preserved as crossover points.

An example of a geometry preserving structure is the helix, Fig. 3.3, the basic structure for yarns, cables and ropes; the deformation of the structure can be quantified by changes in the helix parameters. If as a result of a deformation the helix degenerates into a spiral then the structure is not geometry preserving. To proceed further, it is necessary to identify various nomenclature and parameters; first the *structure,* a rope, shown in Fig. 3.4 is a helix assembly and secondly the component is the element that is formed into a helical path.

Using the parent/child analogy, the component is a fibre in a yarn (structure), or is a strand in a rope. Since in a conventional large rope, the components are strands or subropes which are themselves, parents or structures of ropeyarns or multiplies, the concept of a hierarchy becomes very important. It remains to identify the behaviour of the component within a structure; once these behaviours are

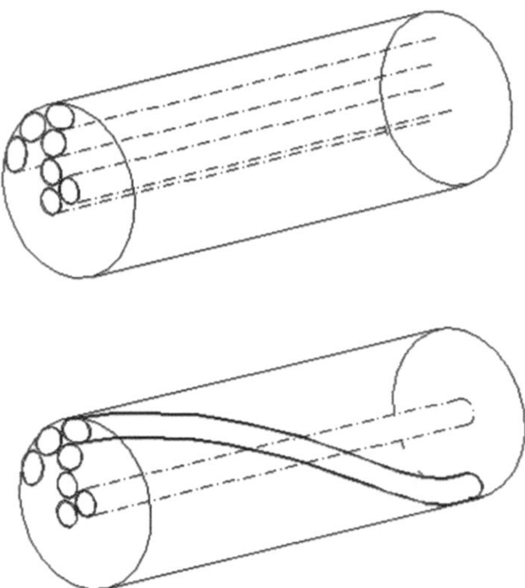

**Fig. 3.3** The helical structure

**Fig. 3.4** A rope structure showing the components (strands and yarns)

determined they can be applied to the different hierarchical relations, that is to fibres within a yarn, to yarns within the strand and to strands within a rope. It is thus very important to determine *'ab initio'* these behaviours for the candidate assembly geometries.

## 3.2 The Helical Structure

### *3.2.1 The Helix Geometry*

Figure 3.5 shows a yarn structure, a single representative fibre in that structure. Also shown is the development from the helix formed by slitting axially, along the surface containing that fibre and opening the structure to form a flat surface. The slit is shown by the Left and Right markers both in the yarn configuration and in the development. In this development the component has become straight. The radius of the helix surface, containing the component is $r$ and the pitch or turns per unit length is $p$; the pitch length $L = 1/p$. The length of the component within a structure pitch length is $l$ and the angle $\theta$ made by the component axis with the structure axis is $\cos^{-1}(L/l)$; in terms of the pitch and helix radius,

$$\cos\theta = \frac{L}{l} = \frac{1}{\sqrt{\left(1 + (2\pi pr)^2\right)}}$$

If the pitch is changed due to stretching and twisting the structure, the attitude of the component (its' direction cosine) is adjusted according to the above relation;

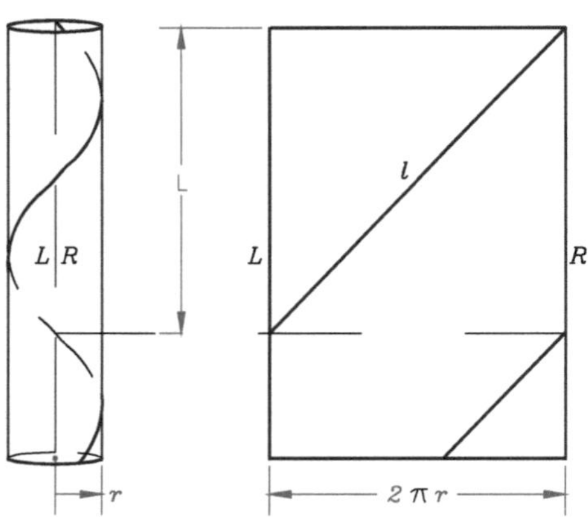


**Fig. 3.5** The yarn structure and helix developmemt

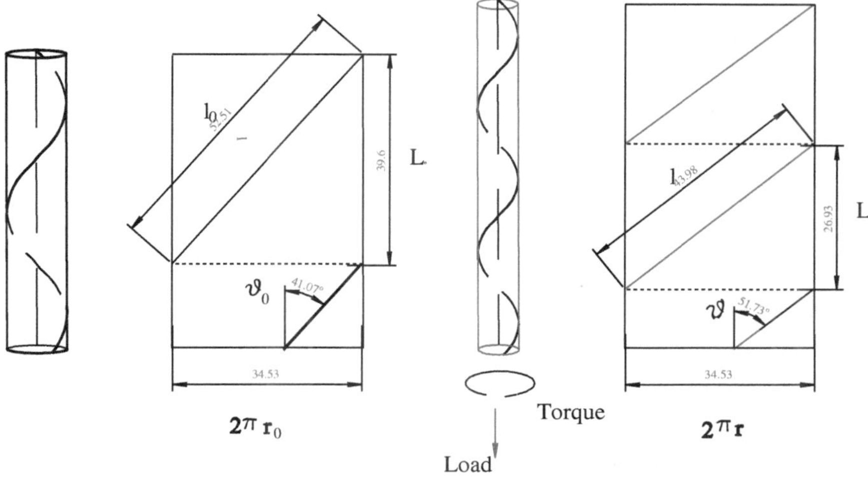


**Fig. 3.6** The helix development and its' deformation

similarly the attitude changes if there is any dilation of the helix by changing the helix radius, Fig. 3.6

The change in helix dimension and twist also results in a change in component length and is as follows

$$l = \sqrt{\left(L^2 + (2\pi r)^2\right)}$$

.

## 3.2.2 Extensional and Twist Deformations

Introducing the structure strain $\varepsilon_s$, the component strain $\varepsilon_c$ is given as follows

$$\varepsilon_c = (1 + \varepsilon_s)\sqrt{\left(\frac{1 + q^2}{1 + {q_0}^2}\right)} - 1$$

where $q = {}^{2\pi pr}/_{1 + \varepsilon_s} \mathit{and}\ q_0 = 2\pi p_0 r_0$ and where the subscript o denotes the reference or initial (undeformed) configuration.

In terms of the direction cosines, the component strain is

$$\varepsilon_c = (1 + \varepsilon_s)\left(\frac{\cos\theta_0}{\cos\theta}\right) - 1$$

The significance of the direction cosines is now apparent, their ratio giving the *advantage* of component to structure strain.

The component strain energy per unit length (J/m) due to component stretching, is $U_e = {}^1\!/_2 EA\varepsilon_c^2$.

If the structure is twisted, the twist per unit length imparted to the component is obtained in two steps; firstly the twist in the structure as a result of the angular rotation of the structure about its own axis induces a twist in the component, $\Phi_c = \Phi_s \cos\theta$ where $\Phi$ is the twist angle and the subscripts $s$ and $c$ denote the structure and component. Secondly the twist/unit length of the component is modified since the component length is increased due to helix twist, and this results in a deformation measure that leads to the torque, as follows $\left(\frac{d\Phi}{dz}\right)_c = \left(\frac{d\Phi}{dz}\right)_s \cos^2\theta$,

since the structure length is $\cos\theta$ times the component length. This expression for twist deformation is also developable from the Frenet-Serret theory, Sect. 3.1.1, that is $\tau = -\underline{p} \cdot \frac{db}{ds} = 2\pi p \cos^2\theta$.

The torsion strain energy per unit component length is $U_t = {}^1\!/_2 GJ\left(\frac{d\Phi}{dz}\right)_c^2 = {}^1\!/_2 GJ(2\pi p)^2 \cos^4\theta$.

### 3.2.3 Flexure in Helical Components

The effect of bending a component into a helix will be introduced by considering the strain energy due to flexure; to obtain this, the component curvature within the helical path must be established and to define this it is useful to use the vector representation of a helix.

Consider a right hand set of orthogonal axis aligned locally with the helix axis; (Fig. 3.7).

The component location is $\underline{r} = r\cos\left(\frac{2\pi s}{l}\right)\underline{i} + r\sin\left(\frac{2\pi s}{l}\right)\underline{j} + \frac{L}{l}s\underline{k}$, where $s$ is the running coordinate along the component, $l$ is the component length in a helix pitch length $L$, and $r$ is the helix radius. Referring to a previous Sect. 3.1.1, General Path Geometry,

the unit tangent is $\underline{u} = \frac{dr}{ds} - \left(\frac{2\pi r}{l}\right)\left(\sin\left(\frac{2\pi s}{l}\right)\underline{i} - \cos\left(\frac{2\pi s}{l}\right)\underline{j}\right) + \frac{L}{l}\underline{k}$,

the principal normal $\underline{p} = \frac{1}{\chi}\frac{du}{ds} = -\cos\left(\frac{2\pi s}{l}\right)\underline{i} - \sin\left(\frac{2\pi s}{l}\right)\underline{j}$ and

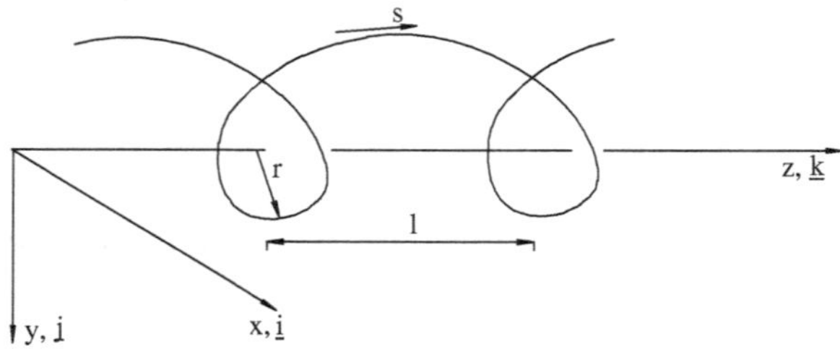


**Fig. 3.7** The helix geometry

the binormal, $\underline{b} = \underline{u} \wedge \underline{p} = \left(\frac{L}{l}\right)\left(\sin\left(\frac{2\pi s}{l}\right)\underline{i} - \cos\left(\frac{2\pi s}{l}\right)\underline{j}\right) + \frac{2\pi r}{l}\underline{k}$

Then $\frac{d^2 r}{ds^2} = -r\left(\frac{2\pi}{l}\right)^2\left[\cos\left(\frac{2\pi s}{l}\right)\underline{i} + \sin\left(\frac{2\pi s}{l}\right)\underline{j}\right]$

and recalling the direction cosine (and the associated sine equation) $\cos\theta = {}^{L}\!/_{l}$ and $\sin\theta = {}^{2\pi r}\!/_{l}$ gives the component curvature $\chi = \left|\frac{d^2 r}{ds^2}\right| = \frac{\sin^2\theta}{r}$.

The strain energy resulting from change in curvature of a component is

$$U_b = U^*(\chi - \chi_0)\frac{l_0}{L_0}$$

where $U^*$ is the strain energy due to bending the component from $\chi_o$ to $\chi$ per unit component length, and $U_b$ is the resulting strain energy per unit structure length.

The forces and torques necessary to cause and sustain the helix deformation against component bending are determined by the Principle of Virtual Work, and are as follows

$$F\delta L + T\delta\phi = L_0\frac{\partial U_b}{\partial\chi}\frac{\partial\chi}{\partial\theta}\left(\frac{\partial\theta}{\partial L}\delta L + \frac{\partial\theta}{\partial\phi}\delta\phi\right).$$

Finally it is pointed out that the change of curvature deduced is not fixed to any transverse axis in the component; in fact to achieve this change in curvature from $\chi_o$ to $\chi$, the initial flexure axis could be deflexed and $\chi$ could be about another axis. The above thus is applicable to a square or circular cross section component using at least a helically isotropic material, Fig. 3.8a.

If a rectangular section is employed there will be a tendency to flex about the axis that is parallel to the long section dimension; this thus remains in the plane of the helix generator and the structure cannot thus be independently torqued, Fig. 3.8b.

### 3.2.4 The Partitioning of Strain Energy

The deformation of a helix by stretching it axially will cause the component to stretch, twist and deflex (by reducing its curvature). The work done in stretching the helix will then be equal to the sum of component extensional strain energy, twist strain energy and bending strain energy. The gradients of these energy with respect to the axial stretch is then a measure of the axial force required against component tension, torsion and bending.

The following three graphs show the strain energy for the three mechanisms plotted against helix strain; the helix radius R, is 0.1 m and component diameter d, is 0.02 m, yielding the dimensionless parameter d/D = 0.1. The initial helix pitch is 0.1 m (10 tpm) for Case 1, 0.01 m (100 tpm) for Case 2 and 0.001 m (1000 tpm) for Case 3, yielding the dimensionless helix parameter (=pR) 1, 10 and 100 respectively. The material stiffness ratio G/E is nominally taken as 0.4, a typical value for steel.

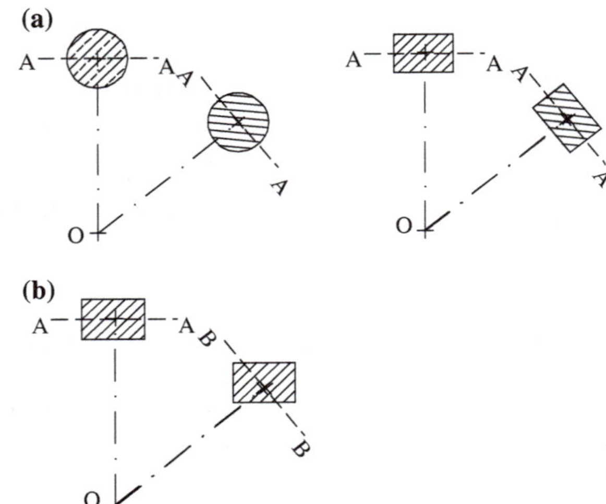


**Fig. 3.8** Flexure about principal and non-principal axes. **a** Cylindrical and rectangluar components(flexed about principal axes, AA). **b** Rectangular component(flexed about non-principal axis, BB)

The graphs show the normalised strain energy for each mode plotted against the helix strain. The strain energy for the secondary modes can be substantially smaller that the extensional strain energy, and are scaled to bring them into the graphical window (Fig. 3.9).

In case 1 the component extensional strain energy dominates; the bending strain energy is substantially larger than the component torsion strain energy (Fig. 3.10).

In the second case where the initial helix twist is 100 tpm, the order of the secondary strain energy is reversed, torsion being much larger than flexural strain energy (Fig. 3.11).

Finally, case 3, the torsion strain energy is larger than that for component stretching which in turn is much larger than that for component bending.

The first case is fairly typical of rope constructions whereas the case 3 is representative of coil springs.

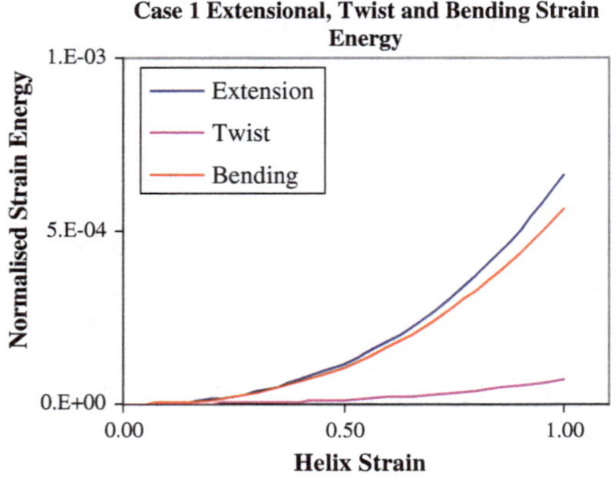


**Fig. 3.9** Strain energy partition for initial helix pitch, 0.1 m (10 tpm)

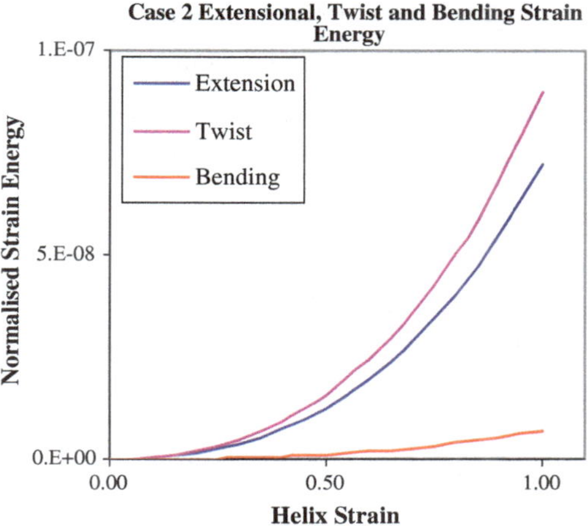


**Fig. 3.10** Strain energy partition for initial helix pitch, 0.01 m (100 tpm)

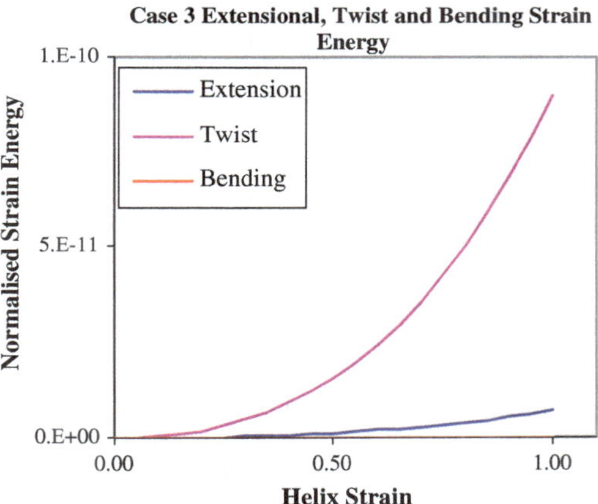


**Fig. 3.11** Strain energy partition for initial helix pitch, 0.001 m (1000 tpm)

The table following summarises the relative magnitude of the component strain energy, by mode due to the helix strain.

| Case | Helix Twist (p) | Dimensionless helix twist (pR) | Twist and Flexural Scale Factor | Energy order |
|---|---|---|---|---|
| 1 | 10 | 1 | 100 | Extensional > Bending > Twist |
| 2 | 100 | 10 | 10 | Extensional > Twist > Bending |
| 3 | 1000 | 100 | 1 | Twist > Extensional > Bending |

The strain energy functionals used in the establishment of the above graphs are

(a) For extension

The component strain energy per unit length (J/m) due to component stretching, is $U_e = {}^1\!/_2 EA\varepsilon_c^2$. Where the component strain $\varepsilon_c = (1 + \varepsilon_s)\left(\cos\theta_0/\cos\theta\right) - 1$ and the direction cosines are $\cos\theta = {}^L\!/_l$ and $\sin\theta = {}^{2\pi r}\!/_l$

For comparison the strain energy is normalised by EA, that is

$$U_e^N = \frac{1}{2}\varepsilon_c^2$$

The significance of the direction cosines is now apparent, their ratio giving the *advantage* of component to structure strain.

(b) For twist

Recalling the expression for torsion, $\tau = -\underline{p} \cdot \frac{db}{ds} = 2\pi p \cos^2\theta$. then torsion strain energy per unit component length is $U_t = {}^1\!/_2 GJ(2\pi)^2(p\cos^2\theta - p_0\cos^2\theta_0)^2$ and normalising

$$\begin{aligned} U_t^N &= \frac{1}{2}\frac{GJ}{EA}(2\pi)^2\left(p\cos^2\theta - p_0\cos^2\theta_0\right)^2 \\ &= \frac{1}{2}\frac{G}{E}\frac{d^2}{8}(2\pi)^2\left(p\cos^2\theta - p_0\cos^2\theta_0\right)^2 \\ &= \frac{1}{4}\frac{G}{E}\pi^2\left((pd)\cos^2\theta - (p_0 d)\cos^2\theta_0\right)^2 \end{aligned}$$

(c) For flexure

Recalling that the component curvature $\chi = \left|\frac{d^2 r}{ds^2}\right| = \frac{\sin^2\theta}{r}$, the flexure energy is $U_b = \frac{1}{2}EI(\chi - \chi_0)^2 = \frac{1}{8}EI\left(\frac{\sin^2\theta}{r} - \frac{\sin^2\theta_0}{r_0}\right)^2$ and the normalised energy is

$$\begin{aligned} U_b^N &= {}^1\!/_2 \frac{EI}{EA}\frac{1}{r^2}\left(sin^2\theta - sin^2\theta_0\right)^2 \\ &= {}^1\!/_8\left(\frac{d}{D}\right)^2\left(sin^2\theta - sin^2\theta_0\right)^2 \end{aligned}$$

## Reference

1. Wempner G (1973) Mechanics of solids with applications to thin bodies. McGraw-Hill, New York

# Chapter 4
# Transversely Continuous Structures

**Abstract** Having dealt with the fibre as the constituent for a structure, linear structures are then examined; these include yarns, cables and ropes. In this chapter, transversely continuous structures are considered; these are structure where the constituent component are so small and so numerous that it is appropriate to consider the structure as a continuum, although in the classis sense it would be anisotropic and inhomogeneous. In this chapter the parallel assembly is also examined and the implications of variability in component length (or prestrain) are quantified.

A linear structure is one with a high aspect ratio; it is effective in tension, is less effective in torsion and is ineffective in flexure. Its function is to position or fix a payload at a prescribed distance from a tether point. Its action is to provide a tensile reaction, a reaction along the axis of the structure whereas a *beam* provides a transverse reaction, normal to the structure axis and a *shaft* provides a torsion reaction around the structure axis. Since fibres have a high aspect ratio and are active in tension, it is logical to use these for components in the linear structure; however because of the small transverse dimension and hence bearing area the load capacity of a fibre is small and to provide a sensible structure, a substantial number of fibres in parallel must be used.

Probably the commonest linear fibre structures are yarns, cables and ropes; they range in size (or diameter) from parts of a millimetre to parts of a metre. Yarns have been extensively studied [1–10] and the theory of yarn mechanics is well established. Their exploitation ranges from use in the fabrication of textile structures to application in tethering, towing and lifting large payloads both dynamically and statically.

In this chapter the geometrical and mechanistic aspects of fibres assembled into linear parallel structures is discussed [1–6]; in the first place the relation between the component, the fibre and the structure (formed from a discrete number of fibres and in a well defined geometry). This establishes the parent/child relationship and forms a basis for the hierarchical tree structure that is essential for the modelling of large rope structures. Because of the complexity of the cross section, it is more sensible to use an energy principle, *Principle of Virtual Work (PVW)* for static analyses and *Hamilton's Principle* for dynamics.

C. M. Leech, *The Modelling and Analysis of the Mechanics of Ropes*,
Solid Mechanics and Its Applications 209, DOI: 10.1007/978-94-007-7841-2_4,

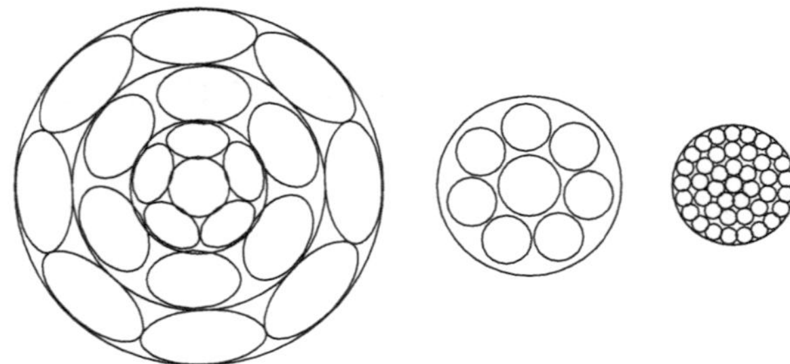

**Fig. 4.1** Cross-section of various assemblies

## 4.1 Geometrically Preserving Structures

To begin the analysis of geometry preserving structures it is useful to consider structures with a large fibre count, so large the discreteness of each fibre is insignificant. The analysis assumes that the fibre size is sufficiently small typically $d/D < 0.001$ or approximately more than a million fibres in the structure; thus they can be assumed infinitesimal compared with the structure and that the operations associated with counting or summation is replaced by integration over the cross section. Figure 4.1 shows various cross sections with layered fibres, a small number of component fibres and also a large number of component fibres.

### *4.1.1 The Assembly of Transversely Continuous Structures*

Consider a parallel structure of n fibres or components, each with a cross section area $A_c$; if they are sufficiently small so that they pack together with no voids (*packing ratio,* $P_f = 1$) the total cross section area occupied by these fibres is $nA_c$ and this is the minimum area that can be developed by these components (Fig. 4.2).

If this structure is now given a twist to form for example a yarn, each component presents an area $A_c/\cos\theta$. Since the cross section area of each component is unchanged then

$$n A_c = 2\pi \int_0^r \cos\theta r dr$$

where $r$ is the structure radius.

This expression can be integrated to give

$$\left(\frac{r}{r_{\min}}\right)^2 = 1 + \left(\frac{\pi p\, r_{\min}}{1 + \varepsilon_s}\right)^2$$

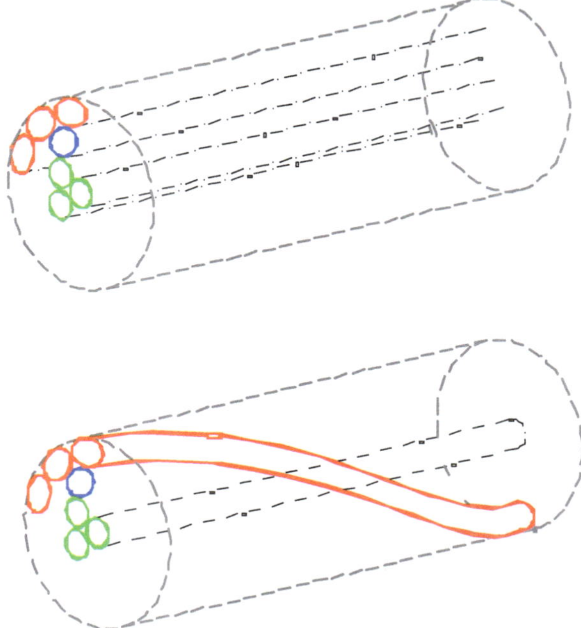

**Fig. 4.2** Parallel and twisted assembly

where $r_{min}$ is the radius of the parallel or minimum area structure, that is

$$r_{\min} = \sqrt{\frac{n A_c}{\pi}}$$

If the initial unstrained structure is formed by twisting a group of these components together, its radius $r_0$ is as follows

$$\left(\frac{r_0}{r_{\min}}\right)^2 = 1 + (\pi p_0 r_{\min})^2$$

Figure 4.3 shows the variation of the structure diameter with system strain for an initial twist, 20 turns/m and an initial structure diameter 1 cm (0.01 m); also shown for the same structure are the diametric changes when the structure is twisted to 30 turns/m and slackened to 10 turns/m. In the former the diameter for zero strain is increased by about 5 % and in the latter it is reduced by 3 %. This figure shows the diametric drawdown due to increasing axial strain and suggests that an effective '*Poisson*' ratio could be estimated; the term '*Poisson*' strictly should not be used here since the Poisson's ratio is a material constant and is applied to a continuum and not a structure. However since the structure is taken to be an aggregate of infinitely small diameter fibres, such a term could be defined for these yarn structures.

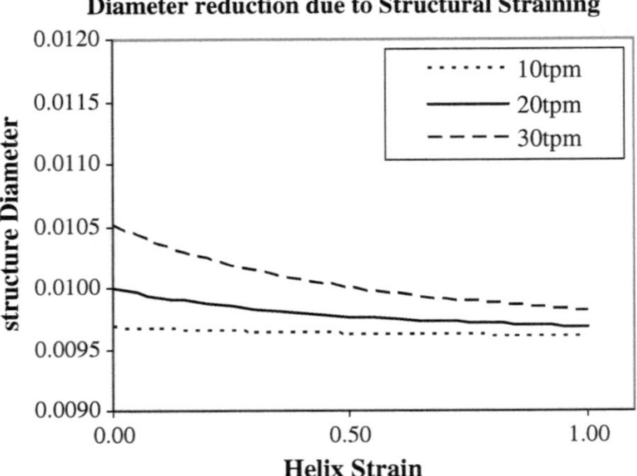


**Fig. 4.3** Diameter reduction due to helix strain

A characteristic length $L_0$ for the structure can be defined from the pitch $p_0$ of the initial structure; a geometric parameter $G$ is then defined as $\pi r_{min}/L_0$ or $\pi p_0\, r_{min}$ and the resulting twisted structure can be assessed from the size of this parameter. For most fibre structures $G \ll 1$, whereas for high twist structures such as coiled springs and some very short pitch wire ropes $G \approx 1$; in this case, the structure is not treated as a continuum since the transverse component diameter is highly skewed, giving rise to assembly problems, necessitating fewer components and hence more voids and a smaller packing ratio. For low twist structures

$$\varepsilon_r = \frac{r}{r_0} - 1 = \sqrt{\frac{1 + \left(\frac{Gp}{p_0(1+\varepsilon_s)}\right)^2}{(1 + G^2)}} - \frac{G^2}{2}\left[1 - \left(\frac{p}{p_0(1 + \varepsilon_s)}\right)^2\right]$$

Figure 4.4 shows the variation of radial strain with the strain and twist of the structure; the twisted structure (0.01 m diameter, 20 tpm, $G = \pi p_0 r_{\min}$) quoted above, is again used.

## *4.1.2 The Straining of Transversely Continuous Structures*

Consider first a parallel assembly of small diameter fibres; if the assembly is given a twist ($p$) in turns/m based upon the unstrained assembly length and strain $\varepsilon_s$, the fibre strain is

$$\varepsilon_c = (1 + \varepsilon_s)\sqrt{1 + \left(\frac{2\pi pr}{1 + \varepsilon_s}\right)^2} - 1$$

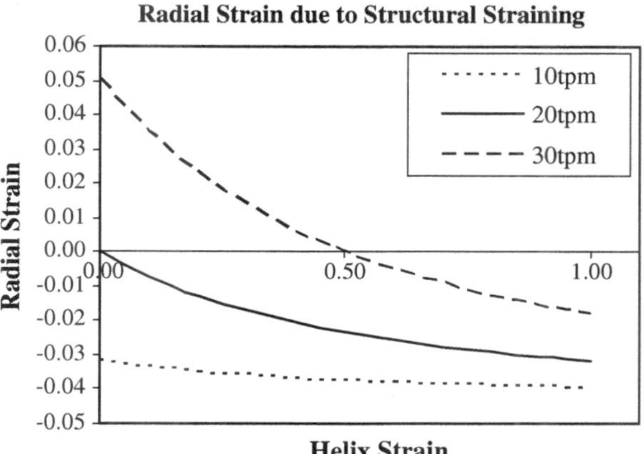


**Fig. 4.4** Variation of radial strain with helix strain

where $r$ is the current radial station of the fibre and is given from the previous section.

Figure 4.5 shows the effect of twisting a parallel bundle of fibres on the strain distribution; the maximum fibre axial strain occurs at the assembly outside, that is where r is maximum (=$R$) and the minimum occurs at the inside where $r = 0$. In this figure it is seen that a twist of 30 tpm on a 1 cm diameter structure causes a 50 % strain in the fibres on the surface for a nominal 10 % axial strain whereas 10 tpm causes 15 % fibre strain at the surface. Failure then occurs at the outside and under further straining and/or twisting it progresses to the interior.

If the initial structure was a twisted structure composed of a set of helically wound fibres the fibre strain in this case is

$$\varepsilon_c = (1 + \varepsilon_s)\sqrt{\frac{1 + f^2 q_0{}^2}{1 + q_0{}^2}} - 1 \quad where \quad q_0 = 2\pi p_0 r_0 \;\; and f = \frac{rp}{r_0 p_0 (1 + \varepsilon_s)}$$

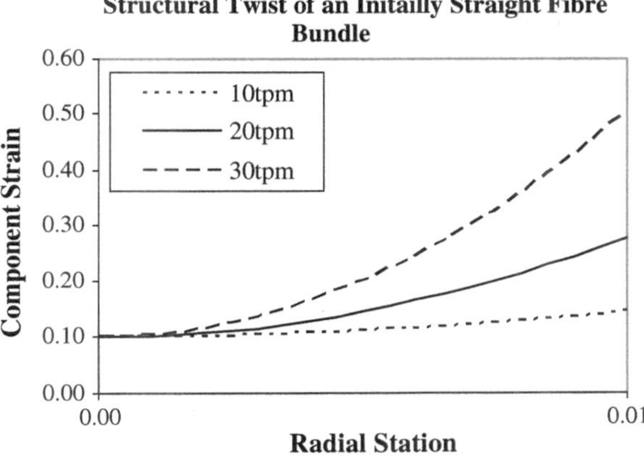


**Fig. 4.5** Distribution of component strain

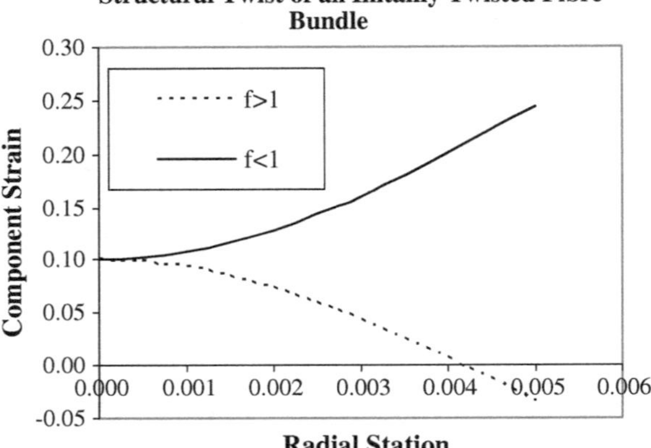


**Fig. 4.6** Distribution of component strain due to pre-twist State

The location of maximum strain now depends on the magnitude of $f$; if $f > 1$ then the maximum strain occurs at the outside and if it exceeds the fibre fracture strain, failure occurs in the outer layers. If $f < 1$ then the maximum strain occurs at minimum $r$ ($= 0$) and failure progresses to the exterior. The radius $r$ and the function $f$ decrease with increasing $\varepsilon_s$.

Figure 4.6 shows the variation of component strain with radial station for $f > 1$ and $f < 1$. Thus if the structure is initially twisted and then strained, there are three possibilities of failure, as follows

(a) if $f < 1$, corresponding to an increased pitch of 30 tpm, the maximum strain occurs at the interior; if this exceeds the critical (failure, yield) strain $\varepsilon_f$ of the fibres then with increasing system strain $\varepsilon_s$, the position where the component strain exceeds the critical strain progresses to the outside, the inner components failing or yielding first. Also note that for larger diameter structures the surface fibres can indeed go into compression.
(b) if $f > 1$, corresponding to a reduced pitch of 10 tpm, then maximum strain occurs at the outside; if this exceeds the critical strain then with increasing system strain $\varepsilon_s$, the position where the component strain exceeds the critical strain progresses from the outside to the inside, the outer fibres failing or yielding first.
(c) if $f$ is marginally greater than 1, again initially the maximum strain occurs at the outside; however since increasing the system strain $\varepsilon_s$ results in a decrease in $f$, there will be a range of radial positioning where due to fibre straightening the fibre strain actually decreases and the critical state now occurs at the inner stations.

### 4.1.3 The Loading of Transversely Continuous Structures

Consider a transversely continuous structure formed from an infinitely large number of very infinitely small diameter components (fibres); the structure is generated by taking these in a parallel bundle and giving the bundle a twist $p_O$ (turns/unit length). The diameter of the structure so formed is $2R_0$. Using the strain of a component located in the undeformed geometry at a radius $r_0$, the component (fibre) strain energy/unit volume is $U^*(\varepsilon_c)$.

The structure strain energy/unit structure length $U_s$ is written as follows

$$U_s = \int_0^{2\pi}\int_0^{R_0} \frac{U^*(\varepsilon_c)}{\cos\theta} r_0\, dr_0 \cos\theta d\psi$$

since the component is longer that the structure, the structure length divided by $\cos\theta$ gives the component length; the second $\cos\theta$ in the integrand is present because the component cross section area is canted with respect to the structure area and the integration represents summation over all the components. Since the structure is axially symmetric then the following results

$$U_s = 2\pi \int_0^{R_0} U^*(\varepsilon_c)\, r_0\, dr_0$$

where $\varepsilon_c$ is a function of the component position $r$ and depends on the helix parameters $p$ and $r$ and the structural strain $\varepsilon_s$; the only parameter that is subject to integration is the radial station $r$.

### 4.1.4 An Analytic Result for Strain Energy

For linear elastic fibres assembled into a low twist structure, there is an analytic result for the structure strain energy; if it is assumed that there is no size effect as a result of straining then

$$\begin{aligned} U_s = \frac{E}{8\pi p_o^2}\Big\{ & (1+\varepsilon_s)^2 (f^2 P + (1-f^2)\ln(1+P)) \\ & - 2(1+\varepsilon_s) < \left(\sqrt{(1+f^2P)(1+P)}\right) \\ & + \frac{(1-f^2)}{f} \ln\left(f\sqrt{1+P} + \sqrt{1+f^2P}\right) + P \Big\} \end{aligned}$$

where $P = (2\pi p_0 R)^2$ and $E$ is Young's modulus of the constituent fibres.

Figure 4.7 shows the variation of strain energy/(unit length)/(unit modulus) of a 1 cm diameter, 20 tpm structure with system strain for different added twists, (10 and 30 tpm); the initial twist structure is strained without twisting and

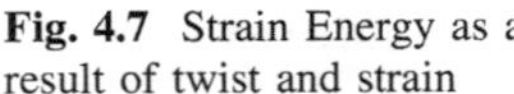

**Fig. 4.7** Strain Energy as a result of twist and strain

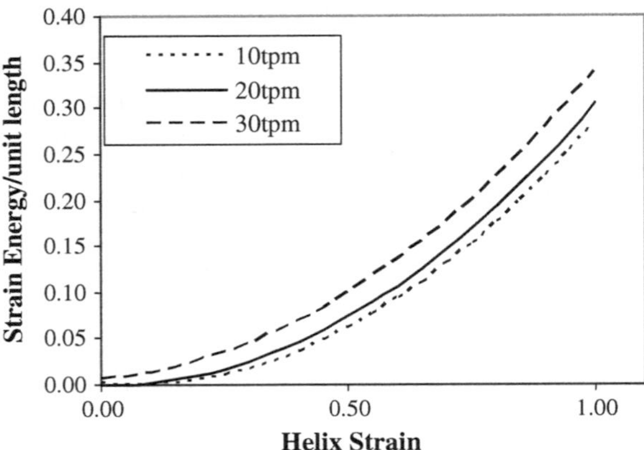


increases fairly monotonically and as expected is seen to be below the dotted curve, the strain energy for a solid rod. If the structure is twisted to 30 tpm prior to straining, it is seen to begin at zero strain with an initial strain energy; if the structure twist is released in part to 10 tpm, it also has an initial strain energy (due to compression of the fibres) which now decreases and then increases as the fibre compressions are removed.

## 4.1.5 *The Principle of Virtual Work for Transversely Continuous Structures*

The *Principle of Virtual Work* uses the structural strain energy and can be applied to give the external force system required to deform the helix and thus the structure; the force components in this first instance are the axial load $F$ and the torque $T$. The virtual work done by these actions is $\delta W = F\delta L + T\delta\phi$ where $\delta L$ and $\delta\varphi$ are the virtual displacements for stretch and twist. The principle of virtual work applied here equates this with the variation of total structure strain energy $\delta U_s$. For a given structure length $L_0$, the principle of virtual work results in the following

$$0 = \delta W - \delta U_s = F\delta L + T\delta\phi - 2\pi L_0 \int_0^{R_0} \delta U^*(\varepsilon_c) r dr$$

The variation of the component strain energy $\delta U^*$ is expressed in terms of the virtual strain $\delta\varepsilon_s$, and this is in turn written in terms of the virtual stretch $\delta L$ and the virtual twist $\delta\psi$, as in the following

$$\delta U^* = \frac{\partial U^*}{\partial \varepsilon_c} \delta \varepsilon_c$$

$$\textit{where } \delta \varepsilon_c = \frac{\partial \varepsilon_c}{\partial \varepsilon_s} \delta \varepsilon_s + \frac{\partial \varepsilon_c}{\partial q} \delta q, \quad \delta \varepsilon_s = \frac{\delta L}{L_0}$$

$$\textit{and } \delta q = 2\pi \left( \frac{\delta(pr)}{(1+\varepsilon_s)} - \frac{pr}{(1+\varepsilon_s)^2} \delta \varepsilon_s \right)$$

The component strain energy above is assumed functionally dependent on the component strain $\varepsilon_c$; it could also be dependent on the component twist and flex in which case there would be added the strain energy due to torsion and bending and these will be discussed later.

The partial derivatives used above are determined and the differentiation needs to be continued to the generalised displacements and is written here for conciseness as

$$F\delta L = 2\pi L_0 \int_0^{R_0} \frac{\partial U^*}{\partial L} r dr \delta L \quad \textit{and} \quad T\delta\phi = 2\pi L_0 \int_0^{R_0} \frac{\partial U^*}{\partial \phi} r dr \delta\phi$$

where the partial derivatives are evaluated as required.

### *4.1.6 The Progressive Failure of Transversely Continuous Structures*

It has been shown that the strain distribution across a transversely continuous twisted structure is non-uniform and depends on the twist parameter, f; if the structure is substantially loaded, the onset of critical component strain occurs at either the core or outside surface and further straining causes more components to be critically loaded. Beyond the critical strain the component will either fail outright (classified as brittle) or sustain a limiting load (ductile and assumed for this to be elastic/perfectly plastic).

In practice, when a component exceeds the fracture strain it will fail locally not totally at all axial stations; its presence will still affect the neighbouring components by inter-component friction. Upon exceeding the fracture strain that component will yield at one (or more discrete locations) and then relax and shed all or part of the load; if the surface contact between components is very smooth, attained by lubrication or by finish, then it could be assumed that component no longer contributes to the load resistance of that structure justifying the brittle model. If however the surface characteristics are such that substantial friction is present, then that component will still experience a sub or near fracture strain since it is not relieved totally by fracture below the fracture load. Here the action of that component in the structure still contributes to the load bearing ability of the structure and can be approximated by an elastic (pre fracture strain)/perfectly plastic (post fracture strain) model.

In each case the variational and integral operators do not commute since the lower or upper limits are subject to variation. To account for this non-commutability, the limits can be kept fixed and the material presence be made a function of the component strain.

#### 4.1.6.1 Brittle Failure

In this case when the fibre strain exceeds the fracture strain, that fibre no longer contributes to the structure strength since the fibre stress is the zero; to account for this the strain energy function is written

$$U_s = 2\pi \int_0^{R_0} \mathrm{H}(\varepsilon_f - \varepsilon_c)\, U^*(\varepsilon_c)\, r_0\, dr_0$$

where $H(\eta)$ is the Heaviside generalised (unit step) function whose value is unit if the argument is greater than zero and is zero if the argument is less than zero; its' value at zero is undefined.

The variation of this is now

$$\delta\, U_s = 2\pi \int_0^{R_0} \left(\mathrm{H}(\varepsilon_f - \varepsilon_c)\delta\, U^*(\varepsilon_c) - \Delta(\varepsilon_f - \varepsilon_c)\, U^*(\varepsilon_c)\delta\, \varepsilon_c\right) r_0\, dr_0$$

where $\Delta(\eta)$ is the Dirac delta function ($= 0$ if $\eta \neq 0$ and is infinite if $\eta = 0$). If $r_{oc}$ is the radius in the reference state at which the fibre strain is critical, that is where $\varepsilon_c = \varepsilon_f$, then the variation of the strain energy now becomes

$$\delta\, U_s = 2\pi \int_0^{R_0} \mathrm{H}(\varepsilon_f - \varepsilon_c)\delta\, U^*(\varepsilon_c)\, r_0\, dr_0 - (2\pi\, r_{oc})\, U^*(\varepsilon_f)\delta\, \varepsilon_c$$

where the variation of $\varepsilon_c$ is evaluated at $r_o$ in the integrand and at $r_{oc}$ in the second term. The structure load can increase beyond that point when the most strained fibres exceed their fracture strain. The structure will usually fail catastrophically.

#### 4.1.6.2 Elastic/Perfectly Plastic Behaviour

In this case when the component strain exceeds the critical (yield) strain, the component still acts but with a constant force; the strain energy function is now

$$U_s = 2\pi \int_0^{2\pi} \left(\mathrm{H}(\varepsilon_f - \varepsilon_c)\, U^*(\varepsilon_c) + (1 + \varepsilon_c - \varepsilon_f)\mathrm{H}(\varepsilon_c - \varepsilon_f)\, U^*(\varepsilon_f)\right) r_0\, dr_0$$

and the variation can again be achieved by accounting for all the terms in the integrand. Again the structure load can increase beyond that state when the most strained fibres exceed their fracture strain and the structure fails progressively.

## 4.2 Parallel Structures, Variability

Probably the simplest linear structure is one composed of many small components assembled parallel to each other and to the structure axis. The advantage of such a structure is that it mirrors precisely the properties of the component in the structure; if there are n components used in the structure, the structure strength is ideally n times the component strength. However there is no scope for equalisation of loads between the components if they are at all unequally assembled. To demonstrate this fault, it is only sufficient to consider an assembly of low breaking strain fibres in a parallel assembly forming a subrope, Fig. 4.8; since the fibres have a low break strain, for example 0.04, any variation of the length of fibres used in the assembly will show up as *preslack* (defined as the minimum system strain at which all components are under tension) when the assembly is under no load and as premature fibre fail when the structure is loaded.

The theory associated with fibre (or component) variability will be based upon a probability distribution since it is not possible to identify any component under any specific prestrain.

Consider a transversely continuous structure, assembled from an infinite number of very thin fibres; the force $F$ in the assembly is given by $F = \int_A \sigma dA$ where $\sigma$ is the component stress, given by $\frac{\sigma}{\sigma_f} = \Im(\eta)$ where $\sigma_f$ is the fracture stress and $\eta$ is the normalised strain ($=\varepsilon/\varepsilon_f$) and where $\varepsilon$ is the fibre strain and $\Im$ denotes a function. Since the stress has been normalised by the fracture stress, the above equation can be used for engineering or specific stresses and indeed in the context of fibre behaviour, the stress can be replaced by fibre tension and fibre breaking load.

Now if there is a maximum preslack $\varepsilon_{ps}$ among the fibres, and defining $\lambda = \varepsilon_{ps}/\varepsilon_f$, the ratio of maximum preslack to component breaking strain, the following change in argument is required

$$\eta - \lambda x = \frac{\varepsilon - \varepsilon_p}{\varepsilon_f} \quad where 0 \leq \varepsilon_p \leq \varepsilon_{ps}$$

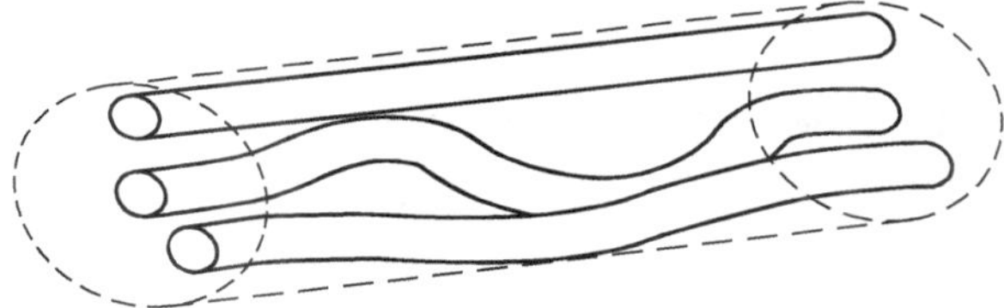

**Fig. 4.8** Parallel component subrope

and where $\varepsilon_p$ is the local preslack and x is a probability coordinate ($=\varepsilon_p/\varepsilon_{ps}$). The probability of occurrence of a constituent with a preslack, $\varepsilon_p$ will vary over the number of components or over the structure according to a probability density distribution, *p(x)*. The probability density function is the probability that a specific value of preslack occurs and is thus a weight attached to that value; it is a requirement that the probability density function satisfies the following integral condition $\int_0^1 p(x)dx = 1$

Then $\frac{F}{A} = \int_0^1 \sigma_f \Im(\eta)p(x)dx$ or $F = \sigma_f A \int_0^1 \Im\left(\frac{\varepsilon}{\varepsilon_f}\right)p(x)dx$

The above integral assumes that all the constituent fibres are acting; in fact they are all acting when $\varepsilon_{ps} < \varepsilon < \varepsilon_f$, range 2. Below $\varepsilon_{ps}$ some fibres are slack (range 1) and above $\varepsilon_f$ some are broken (range 3); for range 1 the upper integration limit is $x = \eta/\lambda$ and the lower limit is 0, for range 2 the upper limit is 1 and the lower limit is 0, and for range 3 the upper limit is 1 and the lower limit is $(\eta - 1)/\lambda$. This applies if $\lambda < 1$ or $\varepsilon_{ps} < \varepsilon_f$ that is where there is a small prestrain variability; if there is a large prestrain variability so that $\varepsilon_{ps} > \varepsilon_f$, implying that some fibres have broken before others have lost their slackness, the upper limit in range 2 is now $(\eta - 1)/\lambda$.

If $F_b$ is the maximum achievable structure load ($=\sigma_f A$) then a variability factor can be defined,

$$\Re = \frac{F}{F_b} = \int_0^1 \Im\left(\frac{\varepsilon}{\varepsilon_f}\right)p(x)dx$$

and this is a function of the local strain ε, and the variability of preslack over the constituent components. The probability density function reflects the manufacturing process is assembling the structure; if it is flat, $p(x) = 1$ then there is equal probability for the occurrence of all preslack within the range. A more useful distribution $p(x) = 6x(1 - x)$ gives a larger probability of prestrain occurrence at the average prestrain; the standard deviation for this function is 0.2236, obtained by integrating $xp(x)$

The above integration could be performed using quadrature algorithms; if the constitutive functional $\Im$ is a polynomial in $\eta$, the integral can be evaluated analytically. To illustrate this, the following assumes a linear elastic fibre law, $\Im(\eta) \equiv \eta$ and the variability factor becomes $\Re = \int(\eta - \lambda x)p(x)dx$.

Using the above probability function, this becomes $\Re = \int(\eta - \lambda x)6x(1 - x)dx$ and can easily be integrated; for range 1 the variability factor is

$$\Re = \eta\left(\frac{\eta}{\lambda}\right)^2\left(1 - \frac{\eta}{2\lambda}\right) \; for \varepsilon < \varepsilon_{ps}$$

For range 2 the variability factor is

$$\Re = \left(\eta - \frac{\lambda}{2}\right) \; for \; \varepsilon_{ps} < \varepsilon < \varepsilon_f$$

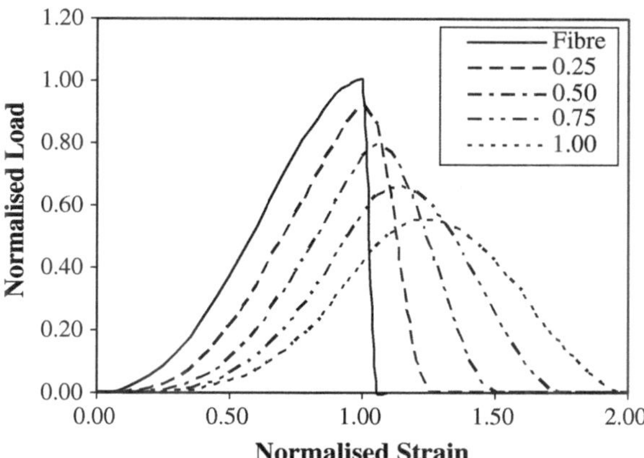


**Fig. 4.9** The effect of variability on an assembly of stiff fibres

and for range 3,

$$\Re = \left(\eta - \frac{\lambda}{2}\right) - \left(\frac{\eta - 1}{\lambda}\right)^2 \left(2 + \eta + \frac{(3 + \eta)(1 - \eta)}{2\lambda}\right) \qquad for\, \varepsilon_f < \varepsilon < \varepsilon_{ps} + \varepsilon_f$$

Figure 4.9 shows the variability factor plotted against system strain for an assembly of stiff fibres, typically Kevlar, breaking strain being 0.04 and for pre-slack of 0, 0.02 and 0.04. The first corresponds to zero variability and is the curve for a 'perfect' assembly of parallel fibres. With variability the curve shows (concave upwards), stiffening as more fibres enter the stressed zone, and lose their preslack. Then follows a linear regime, this being linear since the fibre constitutive behaviour is assumed linear. The final regime is when fibres are broken, those with the largest preslack surviving longest and this part of the curve is softening (concave downwards) and finishes at zero variability factor when all fibres are broken. More variability delays the onset of the maximum load and reduces this, and increases the final break strain. Note that the total work done in completely breaking the system is the area under the curve and is the same for all curves.

## References

1. Treloar LRG (1956) The geometry of multi ply yarns. J Textile Inst 47:T348
2. Riding G (1961) A study of the geometrical structure of multi ply yarns. J Textile Inst 52:T366
3. Frisch-Fay R (1962) Flexible bars. Butterworths, Washington
4. Treloar LRG, Riding G (1963) A theory of the stress strain properties of continuous filament yarns. J Textile Inst 54:T156
5. Kilby WF (1964) The mechanical properties of twisted continuous filament yarns. J Textile Inst 55:T584

6. Riding G (1965) The stress strain properties of multi ply cords, part II: experimental. J Textile Inst 56:T489
7. Treloar LRG (1965) The stress strain properties of multi ply cords, part i: theory. J Textile Inst 56:T477
8. Hearle JWS (1969) On the theory of the mechanics of twisted yarn. J Textile Inst 60:T95
9. Hearle JWS, Grosberg P, Backer S (1969) Structural mechanics of fibers, yarns and fabrics. Wiley, New York
10. Hearle JWS, Thwaites JJ, Amirbayat J (1980) (eds.) Mechanics of flexible fibre assemblies, NATO ASI. Series E No 38, Sijthoff and Noordhoff, German Town

# Chapter 5
# Hierarchical Structures

**Abstract** In this chapter the hierarchical nature of rope structures is identified for the analysis and modelling of ropes and cords and other similar structures; indeed it is indigenous in their manufacture and construction. The various terminology specific to the rope industry is detailed, and the process of structure normalisation is outlined.

The concept of hierarchical structures is described together with the associated candidate assembly processes. The hierarchical tree is described by reference to a three layer rope. The structure (rope) is located at the first level, and its' components (strands or subropes) become the structure at the second level; the third level contains the components to the second level structure (ropeyarns) and these are the structure to the next (fourth) level components, the textile yarns. The final level contains the fibres, filaments or wires, components but not structures as they do not have in this scheme any components. Figure 5.1 demonstrates the hierarchical tree for a three layer (core, inner layer and cover) rope; each layer contains the strands, and within the strands as components are the ropeyarns. The tree shown identifies the fibres in the outer textile yarns of the core rope yarns in the inner layer strands.

There are many other rope (fibre assembly) constructions. Figure 5.2 illustrates a double braid construction enclosing a parallel assembly of ropeyarns.

## 5.1 Component Terminology

Within the rope and textile industries there are various terms that locate the components within the hierarchical ladder [1, 2].

The *fundamental* or starting components used in a linear structure are fibres or filaments and these can be *vegetable* (manila, sisal, coir etc.) or *man-made* (classed as *natural*, e.g. regenerated cellulose, and *synthetic*, e.g. polyamides, polyesters, polypropylene and polyethylene). Also in this latter category are the constituents for wire-ropes namely *wires*. The synthetic fibres can be classified as continuous

C. M. Leech, *The Modelling and Analysis of the Mechanics of Ropes*,

Solid Mechanics and Its Applications 209, DOI: 10.1007/978-94-007-7841-2_5,

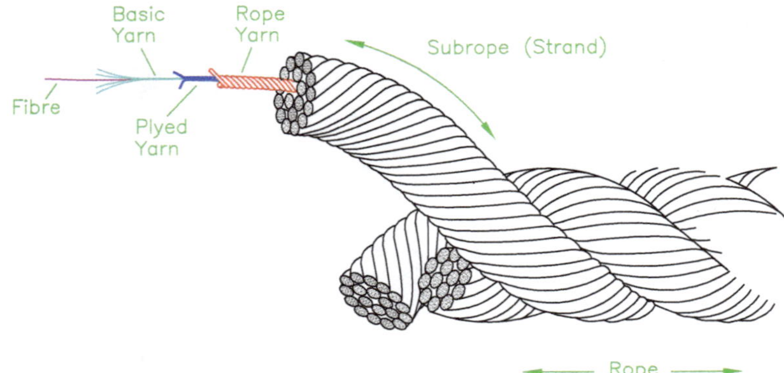


**Fig. 5.1** Typical three strand rope structure

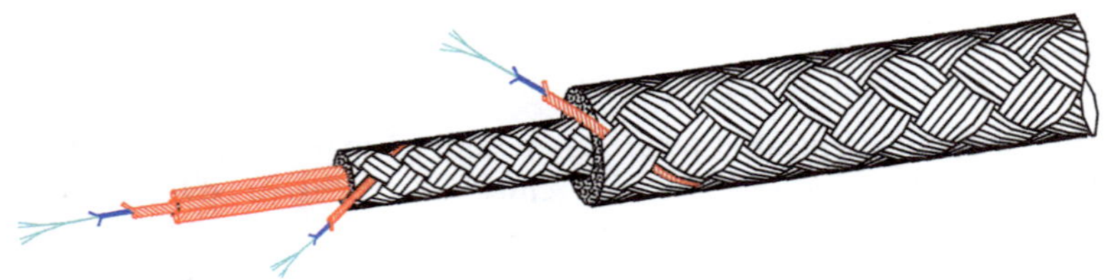

**Fig. 5.2** Hierarchy of a double braid rope

*filaments* where for all useful purposes, they are considered infinitely long or they could be chopped into short lengths, *staple fibres. Monafilaments* are large diameter fibres (typically 0.1 mm) and are stiff or bristle-like; larger diameter fibres (typically 2 mm) are called synthetic wires. Another candidate for the fundamental components is tape chosen for its friction characteristics; this is often shredded longitudinally and is classified as *fibrillated tape* or *film fibres*.

The next level in the hierarchical tree are usually called *yarns* formed by an assembly of the fundamental components (fibres); again various terms are employed to identify this level, these being *yarns*, *baseyarns*, *textileyarns* or *multiplies*. These structures are generated by assembling various fundamental components through a twisting or unusually through a braiding/plaiting process. Sometimes the filaments/fibres are laid together parallel and without twist to form the next level and this could be considered here as twisted with zero twist although other considerations arise from parallel structures.

The next level that can be considered specifically for the cord and rope structures is *ropeyarns* usually formed by a twist grouping of textileyarns.

*Strands* are the next structure in this hierarchical tree and these are formed from an assembly of ropeyarns usually twisted although they could be formed by braiding. The strands used in wire ropes are formed from the fundamental components, namely wires.

*Core*, *subropes*, and *cover or jacket* are components used in the final rope and can be formed from ropeyarns or strands and occur usually in layered ropes, those ropes whose construction is achieved in layers where the inner layer is a core and the outer layer is a braided or extruded jacket.

These cores, subropes, and cover jacket then form a *rope*, *cable*, *line* or *cord* although the rope, cable, line or cord could be assembled from strands or indeed from ropeyarns only.

The above components used in a hierarchical structure are only typical; various structures could be composed from some of the above elements and not necessarily follow the tree structure. For example the 24 mm diameter three strand nylon rope, Fig. 5.3 contains three strands, each strand consists of sixteen ropeyarns. Each ropeyarn consists of four baseyarns or textile yarns; the baseyarns are considered to be the fundamental components for analysis even though they can consist of many continuous nylon fibres.

The assembly process for generating a higher level in the hierarchical tree from lower or smaller components will use a twisting, braiding or plaiting process; these are modelled by one of the assembly models defined previously using assumptions relating to the size of the lower components, their number and their transverse stiffness.

The nature of hierarchy can be classified as *simple* or *compound*; simple structures are those in which the components in a level are *singular in type* although there could be many of them. In the above example, Fig. 5.3, the 24 mm nylon rope is simple at its first level since there are three identical strands; it is also simple at the lowest level where there are four baseyarns, equally disposed in each ropeyarn. It is not simple at the intermediate level, strands to ropeyarns since these are wrapped in layers and each layer will cause a different strain in the ropeyarns in that layer; the ropeyarns will fail at different rope strains, those at the strand core failing first. Each component in a simple structure is equally strained and exerts the same force within the structure; consequently they all fail at the same structure strain and thus a simple structure at least in theory fails catastrophically.

**Fig. 5.3** Three strand rope structure showing the strands and ropeyarns

For example the three strand rope above could have the following disposition of components

| | |
|---|---|
| Level 1 | 3 Stand rope<br>(Homogeneous, compound, half a million fibres) |
| Level 2 | 3 Strands |
| Level 3 | 35 Ropeyarns |
| Level 4 | 4 Textileyarns |
| Level 5 | 1000 Nylon fibres |

A compound structure can be generated by either using different components within the same level, by mixing components within the same layer or by using different components in different layers, or by using a construction that differentially strains the components (in transversely continuous structures or layered structures). Since the directions of the components are different in different layers each layer is subject to different strains and the components will approach failure differently. The failure of such (complex) structures is thus progressive, as each group of components fail, the structure load is shared with those surviving components. Typical of the compound structure are the braids, Fig. 5.2.

| | | | |
|---|---|---|---|
| Level 1 | Braided rope<br>(Homogeneous, compound, 22.64 million fibres) | | |
| Level 2 | Core | Inner Braid | Jacket |
| Level 3 | 32 Parallel subropes | 16L and R strands | 16L and R strands |
| Level 4 | 19 Ropeyarns | 4 Ropeyarns | 4 Ropeyarns |
| Level 5 | 35 Textileyarns | 5 Textileyarns | 5 Textileyarns |
| Level 6 | 1000 Nylon fibres | 1000 Nylon fibres | 1000 Nylon fibres |

A *homogeneous* structure is one in which all the constituents are of the same material; an *inhomogeneous* structure will use different materials in the construction, for example a structure might employ a mix of Kevlar and nylon ropeyarns or a rope could be designed with a Kevlar core (for strength) and a polyester cover (for handling and containment). A simple structure is by definition homogeneous but a homogeneous structure is not necessarily simple, since it could use a mix of different constructions with a level.

## 5.2 Normalisation of Hierarchical Structures

The effect of various assembly geometries and the associated parameters is to produce a structure with a behaviour that is developed from but different to that of its constituents; this modification is usually called the *realisation* of an assembly and can be thought of as an efficiency parameter.

To assess this, the structure is *normalised* by measuring its performance in terms of specific stress; if this is then compared with the specific stress of the components in each hierarchical level, the effect of realisation is quantified. This is shown in graphical form; during a deformation process achieved by straining the structure, the specific stress (component load/component weight per unit length) is plotted against component strain. Realisation reduces the peak specific stress and increases the final breaking strain of the structure.

For a homogeneous structure, the area under the curve (energy to break/unit mass of material) for the structure is the same as that for each component since the energy to break the structure must be the same as that required breaking the constituents.

For an inhomogeneous structure, the area under the structure curve is the same as that under the weighted component curves, each being weighted by its mass presence.

## References

1. Himmelfarb D (1957) The technology of cordage fibres and rope. Leonard Hill (Books) Ltd, London
2. Klust G (1983) Fibre ropes for fishing gear. Fishing News Books, Surrey

# Chapter 6
# Transversely Discrete Structures

**Abstract** This chapter examines transversely discrete structures, those in which the components are easily countable and identifiable. The component (fibre, yarns, strands etc.) elements are assembled by twisting or weaving (braiding, plaiting) into another linear component whose dimensionality (L/d) is reduced from that of the constituent component. Tension and torsion behaviour are the focus, but by implication in the modelling transverse deformation is important is developing the model for the structure behaviour. This arises from the compression and distortion of components as the structure is loaded. Also in this chapter the various component nomenclatures are listed and the various component assembly configurations are described. The hierarchical nature of rope structures is important for the analysis and modelling of ropes and cords and other similar structures; indeed it is indigenous in their manufacture and construction. Finally bending of helical structures is introduced with the two extreme assumptions, no slip and zero friction.

In this chapter the modelling of an assembly of linear components is developed. There is a difference between the modelling of a large number of small discrete components and the modelling of a small number of large components, so that these are considered separately. The modelling of non geometry-preserving structures, typically braided and plaited structures is developed. Finally the bending of these structures is introduced.

## 6.1 Numerous Small Component Structures

In the Chap. 5, it was assumed that for those structures there were so many (small) components that it would not be useful to consider them separately. They could be assumed to be sufficiently small and numerous that the structure could be assumed continuum. In this section a smaller number of larger components is considered, still numerous but distinct, typically yarns consisting of say 50 fibres. The geometry of assembly is now very important. This extrapolation is also taken a

C. M. Leech, *The Modelling and Analysis of the Mechanics of Ropes*, Solid Mechanics and Its Applications 209, DOI: 10.1007/978-94-007-7841-2_6,

step further where a small number of relatively larger components is examined, for example the 3 strands in a three strand rope. In this case the analysis will consider inter component deformations [1–9].

### 6.1.1 The Hierarchical Tree

The structure can then be considered as one of a set of components now used to from a new structure and this is identified as a hierarchical structure in which the final structure is composed of structures and the geometry used to form the final structure is one of a class of geometries that contains that geometry used to form the component structure. This hierarchical concept can be extended both ways to reduce the structure to its simplest components and to yield the ultimate useful structure that can contain half a million tangible fundamental components.

If the number of components in a twisted structure is large but not excessively, the assumption of continuity across the structure is no longer valid; the discreteness and countability of components must be considered.

The structures considered here will have a large number of components but typically this count will be less than a million; these are disposed in a small number of groups, in layers or types. It is these groups or types that enable the structure to be labelled as transversely discrete, although the components in the groups could be transversely continuous. The assembly would be achieved by taking the large number of components, $N$ in a parallel configuration as if they had been combed axially, and then to give the structure a twist ($p$ turns/m) under no load; in this twisted state each component is *stress free.*

The geometry, Fig. 6.1, is a multilayer assembly of components, diameter d. The construction is such that one component at the core is twisted about its own axis; however since the aspect ratio is small the effect of this fibre torsion can be and is often assumed negligible. Outside this core and in the next layer there are no more than 6 components and these act identically and form the first layer. Outside this layer there are no more than 12 components, again acting identically and these form the second layer. This progression continues until all the components have been incorporated into the structure.

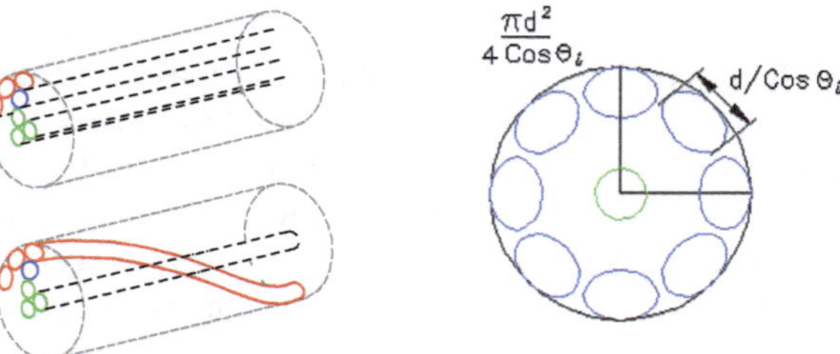


**Fig. 6.1** Schematic representation of layered twisted assembly

The assumption that all in a layer act identically is not always realistic, especially if the components have a low breaking strain and a high modulus; in this case the variability theory developed in the Chap. 5 would be applied to the components within the layer.

### 6.1.2 The Assembly of Transversely Discrete Structures

The helix can be analysed by cutting the surface along a line on the surface, parallel to the central line, Fig. 6.2.

The direction cosine for the components in the *i*th layer can be determined by the following

$$\cos\theta_i = \frac{1}{\sqrt{1+(2\pi pr)^2}} = \frac{1}{\sqrt{1+(2)^2}}$$

The fitting of components into the *i*th layer is $n \le 2\pi i\cos\theta_i$, this resulting from the fitting of components into the circumference in the middle of the *i*th layer; however there is also a packing criterion, specifically that the space occupied by the components fit into the layer space, $n \le 8\,P_r\,i\cos\theta_i$ where $P_r$ is the packing ratio; the total number of components $N$ within all the total layers is then

$$N = 1 + \sum_{i=1}^{l} n_i = 1 + \min(2\pi, 8\,P_r)\sum_{i=1}^{l} i\cos\theta_i = 1 + \min(2\pi, 8\,P_r)\frac{l(l+1)}{2}\cos\theta_a$$

where $l$ is the number of layers in the structure and $\cos\theta_a$ is the average direction cosine of the components in the structure.

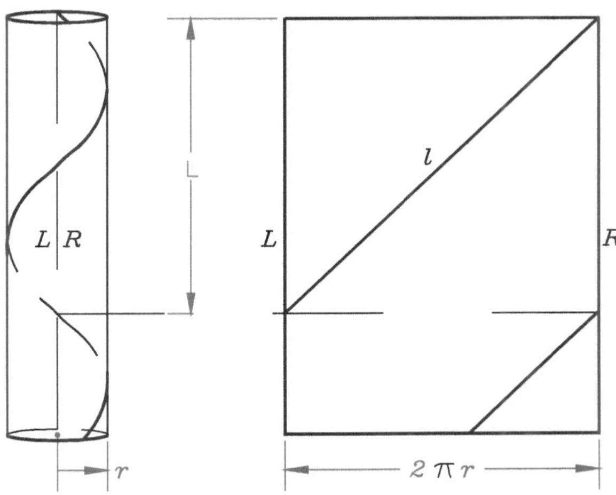


**Fig. 6.2** The helix development

### 6.1.3 The Straining of Transversely Discrete Structures

Consider the *ith* layer formed from $n_i$ small diameter components (fibres); this layer is formed by taking these $n_i$ components and wrapping them in parallel around the inner (*i*-1th) layer so that they each describe a helix with a twist $p_0$(turns/unit length) (Fig. 6.3).

The diameter of the helix or centre of that layer is *id* or 2*ir*. The strain of a component has been developed and is as follows

$$\varepsilon_{ci} = (1+\varepsilon_s)\sqrt{\left(\frac{1+{q_i}^2}{1+{q_0}^2}\right)} - 1$$

where

$$q_i = \frac{2\pi p\, r_i}{1+\varepsilon_s} \quad \text{and} \quad q_0 = 2\pi p_0\, r_0$$

and hence the component(fibre) strain energy/unit volume, $U^*(\varepsilon_{ci})$ can be quantified.

The structure layer strain energy/unit structure length $U_i$ is written as follows;

$$U_i = n_i\, U^*(\varepsilon_{c_i})\frac{\pi\, d^2}{4}$$

where $\varepsilon_{ci}$ is the strain of the components in that layer and depends on the helix parameters p and r and the structural strain $\varepsilon_s$; each component in that layer is assumed to behave identically since there is no suggestion of variability within a layer; this aspect is considered later.

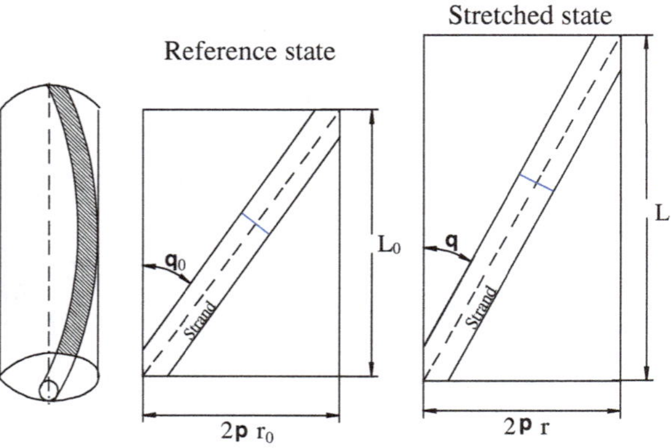


**Fig. 6.3** The helix development, with the reference and strained states

### 6.1.4 The Principle of Virtual Work for Transversely Discrete Structures

The *Principle of Virtual Work* has been outlined earlier. Here the application to an assembly of layers, each layer consisting of a discrete number of components; the total structural strain energy is obtained by summation over all layers. The variation of the total structural strain energy then is equated to the virtual work done by system strain and twist. This is applied to give the external force system required to deform the helix and thus the structure; the force components in this first instance are the axial load $F$ and the torque $T$. The virtual work done by these external actions is $\delta W = F\delta L + T\delta\phi$ where $\delta L$ and $\delta\varphi$ are the virtual displacements for stretch and twist. The principle of virtual work applied here equates this with the variation of strain energy $\delta U_s$.

For a given structure length $L_0$, this results in the following

$$0 = \delta W - \delta U_s = F\delta L + T\delta\phi - L_0 \sum_{i=0}^{layers} \delta U_i$$

The variation of the structure strain energy $\delta U_s$ is expressed in terms of the virtual strain $\delta\varepsilon_s$, and this is in turn written in terms of the virtual stretch $\delta L$ and the virtual twist $\delta\psi$, as in the following

$$\delta U_i =_i \frac{\partial U_i^*}{\partial \varepsilon_c} \delta \varepsilon_c$$

$$\text{where} \quad \delta\varepsilon_c = \frac{\partial \varepsilon_c}{\partial \varepsilon_s}\delta\varepsilon_s + \frac{\partial \varepsilon_c}{\partial q}\delta q, \;\; \delta\varepsilon_s = \frac{\delta L}{L_0} \quad \text{and}$$

$$\delta q = 2\pi\left(\frac{\delta(pr)}{(1+\varepsilon_s)} - \frac{pr}{(1+\varepsilon_s)^2}\delta\varepsilon_s\right)$$

The component strain energy above is functionally dependent on the component strain $\varepsilon_c$; it could also be dependent on the component twist and flex in which case there would be added the strain energy due to torsion and bending. These will be discussed later.

The differentiation needs to be continued to the generalised displacements and is written here for conciseness as

$$F\delta L = L_0 \sum_{i=0}^{layers} \frac{\partial U_i}{\partial L}\delta L \quad \text{and} \quad T\delta\phi = L_0 \sum_{i=0}^{layers} \frac{\partial U_i}{\partial \phi}\delta\phi$$

## 6.2 Component Geometries

In the previous sections, it has been assumed that components are numerous and also transversely stiff; the first assumption is appropriate for many yarn type structures but for rope and some strand structures the number of components is small, typically less than 10 and their size is relatively large so that the component diameter to structure diameter, the constructional dimensionality *(d/D)* is probably greater than 0.1. It cannot be assumed now that there is no contact between neighbouring components in the same layer; in fact the converse is an essential for the structure. The component is now suspended and restrained from migrating to the core of the structure not by the presence of material in the path between it and the core, but by its' neighbours acting on the component sides or shoulders. This resistance is converted to a shear force which would distort the cross section of the component and by normal force acting toward the component centre, Fig. 6.4.

These contact forces act initially on a small contact zone between the contiguous components; depending on the interior stiffness of the component, these actions will cause distortion and dilation resulting in a change in cross section shape and area. Two extremes are examined here, firstly when the component is transversely stiff or hard so that there is essentially no change in cross section due to these shoulder forces and the other extreme when the component is soft so that the shoulder forces cause the component to fit completely into the available area.

The first would be a suitable assumption for wire rope strands in a rope, and the second is appropriate for deformation in the conventional synthetic and natural fibre ropes. Some fibre rope strands are transversely stiff and then neither assumption is appropriate; however the limits generated by their consideration would contain most cases.

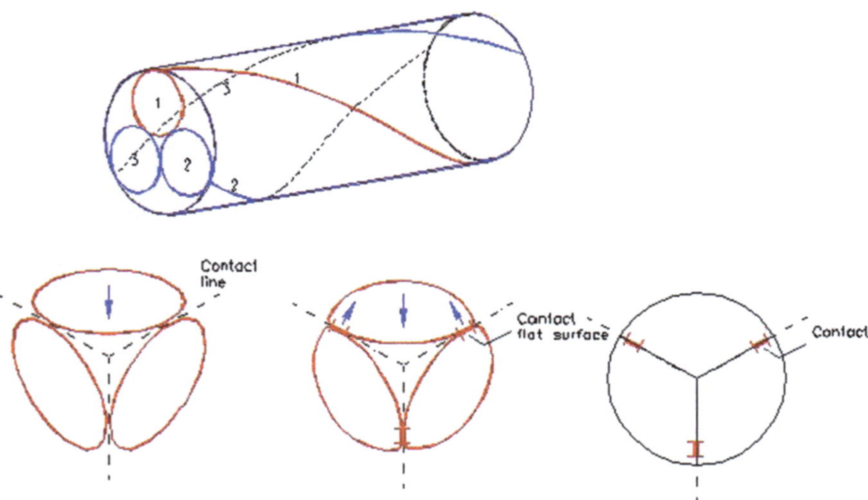


**Fig. 6.4** The deformation of components within a structure under load

### 6.2.1 The Assembly of Large Component Structures

The geometry considered here is composed of a small number of components, circular in cross section, disposed uniformly around the structure centre and standing off from the structure axis; consider now n components each with a diameter $d$, located at a radius $r$ from the structure centre and each component follows a helix about the structure axis with p turns/unit length. The component attitude or direction cosine is related to the pitch and the helix geometry by the following

$$\cos\theta = \frac{1}{\sqrt{1 + (2\pi pr)^2}}$$

The following development depends on the assumptions used concerning the component transverse stiffness; hard and soft components are considered, the former where the component is rigid in the transverse direction and the latter where the component is totally yielding but also incompressible.

a. Transversely hard components

From the helix parameters, each component presents an approximation to an ellipse in the structure cross section; the ellipse minor axis, $d$ is aligned with the structure radius and the major axis, $d/\cos\theta$ is aligned along the structure circumference (Fig. 6.5).

Since each of the $n$ components subtends an included semi-angle $\psi(= \pi/n)$ at the structure axis. The component must fit into this included angle and thus the contact occurs where the tangent to the ellipse is at $\psi$ to the ellipse minor axis. The connection between the helix radius $r$ and the number of components and the component attitude ($\theta$) is given as follows

$$\frac{2r}{d} = \sqrt{1 + \cot^2(\psi)\sec^2\theta}$$

From this and the helix equation, the direction cosine can be determined in terms of the various fixed parameters, specifically the component diameter, the component number and the helix pitch as follows

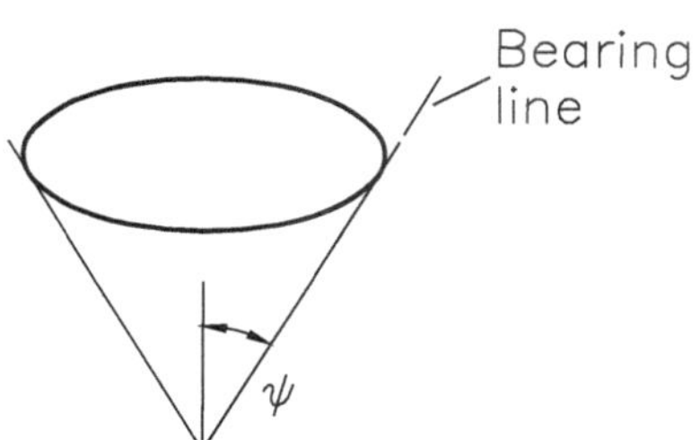


**Fig. 6.5** The hard component

$$\cos\theta = \sqrt{\frac{1 - \cot^2(\psi)(\pi pd)^2}{1 + (\pi pd)^2}}$$

and the resulting helix radius can then be obtained. These equations are limited two-fold; obviously $n$ must be an integer and larger than 2, ($n \geq 3$ or $\psi \leq \pi/3$). Secondly there is a fitting or assembly criterion which is governed by the number of components, their size and by the number of turns/unit length, $p$; this criterion determines whether for a given pitch and size, the structure can be formed and this is expressed as

$$\tan\psi \geq \pi pd \quad \text{or} \quad n \leq \frac{\pi}{\tan^{-1}\pi pd},$$

Referring back to the principle of virtual work, it can be seen that in the above $\cos\theta$ is a function of the pitch length $p$; thus the variation of component strain energy can be written in terms of a variation in component strain which is expressed in terms of variations in structure strain and direction cosine and hence in variation of structure strain and pitch length (Fig. 6.6).

b. Transversely soft components

For soft components the components can be squashed to fit into the available sector; the component is seen to migrate to the structure axis and forms a sector. This is an assumption, the real component shape being somewhere between an ellipse and the sector. For reasonable modelling the sector approximation will be taken as the alternate extreme to the hard component (stand-off) geometry (Fig. 6.7).

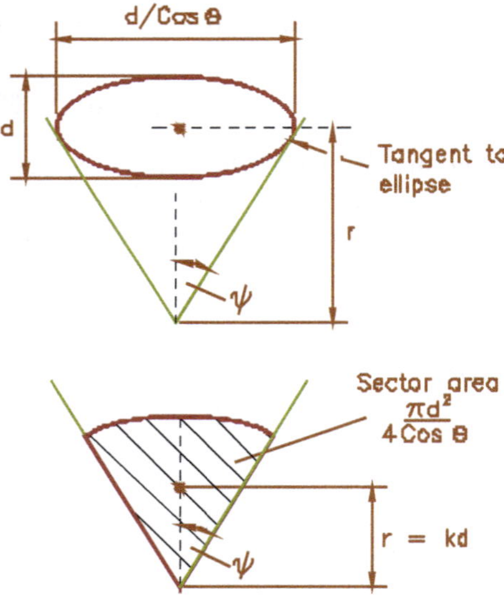


**Fig. 6.6** Component geometry for hard and soft configurations

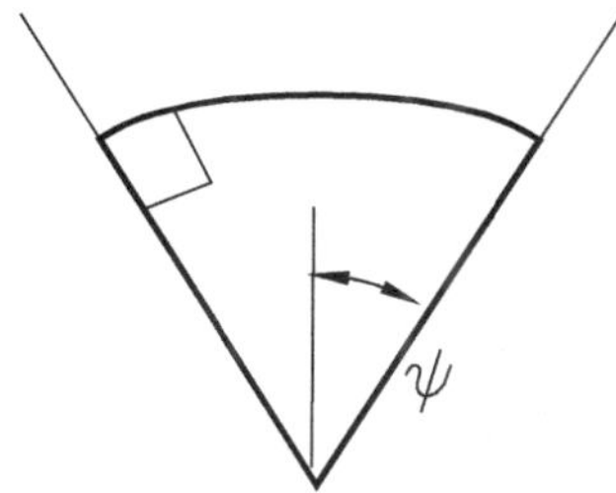


**Fig. 6.7** The soft component, fully deformed

The area of the component presented to the structure axis is $A_c(=\pi d^2/(4\cos\,\theta))$ and this is fitted into the available sector as follows

$$A_c = \frac{\pi\, d^2}{4\cos\,\theta} = R^2\,\psi$$

Since the helix radius is related to the sector outside radius $R$ by $r = \kappa R$ where $\kappa\#1$, then it is possible to evaluate the direction cosine as in the following

$$\cos^2\theta + \beta\cos\theta - 1 = 0$$

$$\text{where } \beta = (2\pi p\kappa)^2 \frac{\pi\, d^2}{4\psi} = n(\pi p d\kappa)^2$$

The value of the parameter $\kappa$ is an assumption; it could be the location of the sector centroid in which case $\kappa = \frac{2n\sin\pi/n}{3\pi}$ or it could be the geometric centre of the sector so that $\kappa = 0.5$; the first assumption implies that the centre of area and the centre of twist are coincident whereas the second implies a geometric centre fixed by the radius of the sub-components within this component.

The specific choice for the centre of twist would need to be justified but there is very little difference between these candidates, and these are certainly within engineering and textile approximations.

c. The transition from hard to soft components

Assume that in the assembly process there are n large diameter components twisted around a common axis; initially neighbouring components will be in mutual contact along a line that runs parallel to the component axis (a) (Fig. 6.8).

Under a structure load, which is developed from the component loads, this contact line will be forced into a bearing surface (b); if the components are elastically stiff then an approximation to this behaviour could be developed from Hertzian theory. That part of the component not in contact and initially elliptical to the structure axis will form a new curves; an assumption that is useful is to assume that the projected area is constant and that the two free edges are ellipses tangent to the contact surfaces, (c). Increasing the structure load will increase the contact area and will bring the inner ellipse close to the structure axis; at the limit the contact area will stretch from the outer free surface to the structure axis, (d).

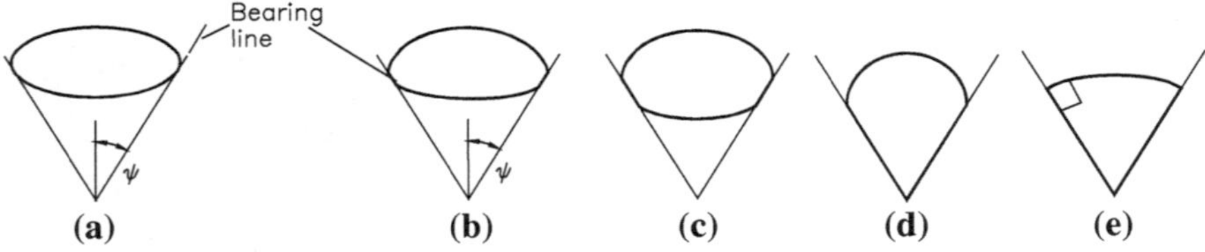


**Fig. 6.8** Component distortion

A further increase in load will bring the component material closer to the structure axis and this will be achieved by changing the outer ellipse, tangent to the contact line to an arc centred at the structure axis, (e); this is accomplished by increasing the angle made by the free surface and the contact surface from zero through to $\pi/2$ due to component drawdown, (wedging).

A possible mechanism for the change of shape from elliptical (really circular) to a wedge is shown in Fig. 6.9. The central components however must stay on the centre line and can only get closed together and smaller in size.

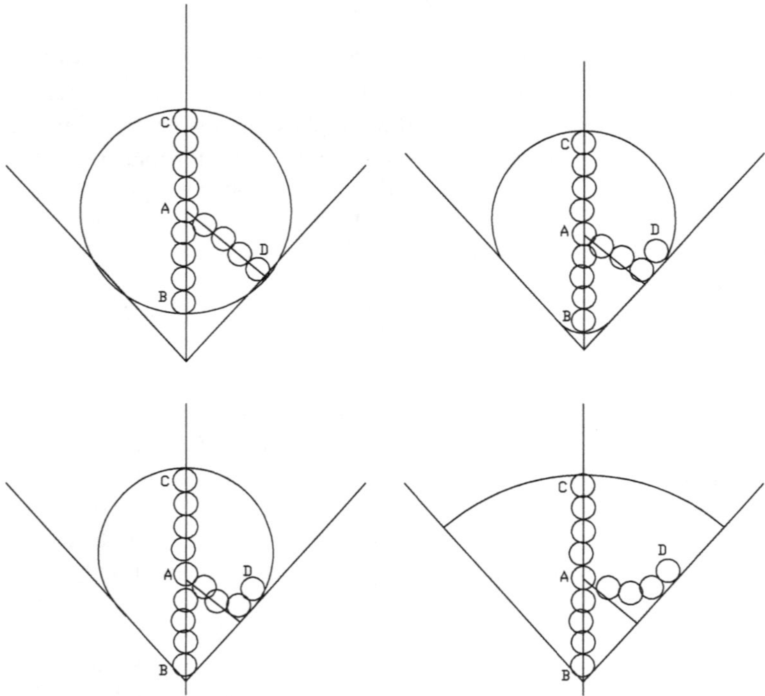


**Fig. 6.9** A possible detail for sub-component migration

### 6.2.2 The Loading of Large Component Structures

As in the previous section where loading of transversely discrete structures is considered, the conversion of structure strain to component strain is the essential concept. The diameter of the helix is given above for soft and hard components; this varies with the structure strain, the structure twist and the formulae are restated here in the form useful to large components, as follows

$$\varepsilon_c = (1 + \varepsilon_s)\sqrt{\left(\frac{1 + q^2}{1 + {q_0}^2}\right)} - 1$$

where

$$q = \frac{2\pi pr}{1 + \varepsilon_s} \quad \text{and} \quad q_0 = 2\pi p_0 r_0$$

Again the component (fibre) strain energy/unit volume, $U^*(\varepsilon_c)$ can be quantified. The structure strain energy/unit structure length $U_s$ is written as follows:-

$$U_s = nU^*(\varepsilon_c)\frac{\pi d^2}{4}$$

where $\varepsilon_c$ is the strain of each of the n components and this depends on the helix parameters $p$ and $r$ and the structural strain $\varepsilon_s$; each component is assumed to behave identically since there is no variability within that group.

Figure 6.10 shows the load-extension behaviour of a 3 strand nylon rope; the ordinate is load/weight (dN/Tex) since on this scale it is possible to compare the structure along with the components and the abscissa is strain.

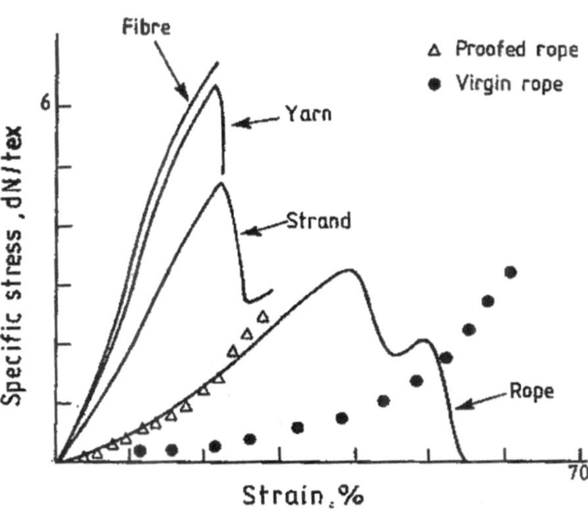


**Fig. 6.10** The extension of a three strand nylon rope

The figure shows the behaviour for transversely soft components. Also shown are experimental data, the virgin rope where the components are assumed hard, circluar and the proofed, where the components have been worked into the sector configuration. Note that the proofed rope has a larger break load and smaller break strain.. This is because the helix used by the hard, virgin components has a larger radius and hence the direction cosine is smaller; this has the effect of reducing the strain in the components. The components (fibre, yarn, strand) are also shown on the same scale.

## 6.3 Tubular Woven Geometries: Braiding and Plaits

In the braiding process the strands are not tied together by twisting but by interfacing in such a way that they cross each other in a diagonal pattern.

The geometry of tubular braided rope is shown in Fig. 6.11; the components in a layer are divided into two sets, those that wind left, clockwise and those that go right, anticlockwise. Conventionally the components that go right, if strands, have subcomponents, ropeyarns that are laid left and conversely for those strands that go left. have ropeyarns that go right. The amount of empty space in the inside of the tube decreases with the number and the volume of the strands. The detail of the interaction between left ad right strands is shown in Fig. 6.12.

The strands forming the braid are paired in mirror type; those migrating left or 'S' have a complement migrating right or 'Z'. The constitution of the S strands is identical to those Z strands except they are twisted in the opposite sense. The effect is to form a torque balanced structure, balanced at all loads (Fig. 6.13).

Figure 6.14 shows the load extension of double braided rope; also shown are the two strands (core and jacket), yarns and fibres. This shows that structure is relatively stiff, and that there are two strand behaviours. The inside strand, the core breaks first at about 10 % strain and the second, jacket continues beyond this point.

Eight strand braids, plaited ropes have essentially no core space in the centre; the component paths for eight strands are shown in Fig. 6.15.

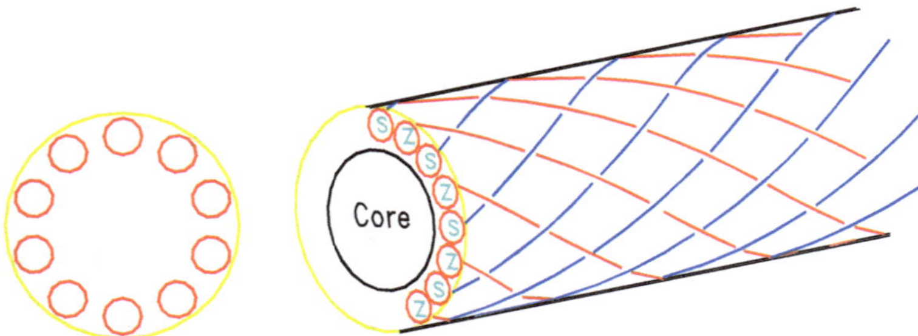


**Fig. 6.11** Tubular braided rope

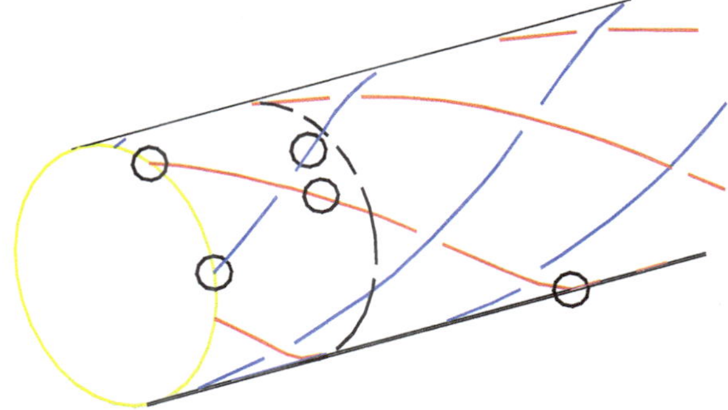

**Fig. 6.12** Detail of braid

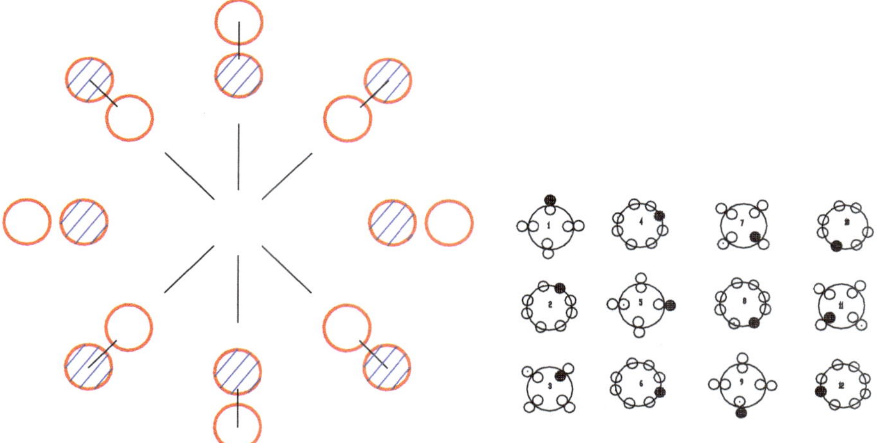

**Fig. 6.13** The cross-section of braided assemblies (*Shaded* components rotate to *left*, *anticlockwise*)

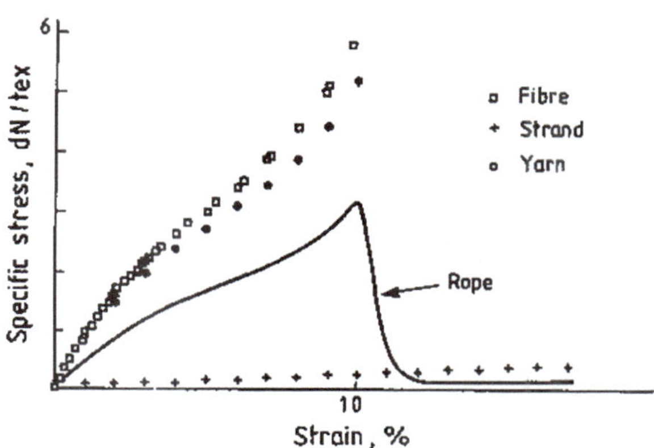


**Fig. 6.14** The extension of a double braided nylon rope

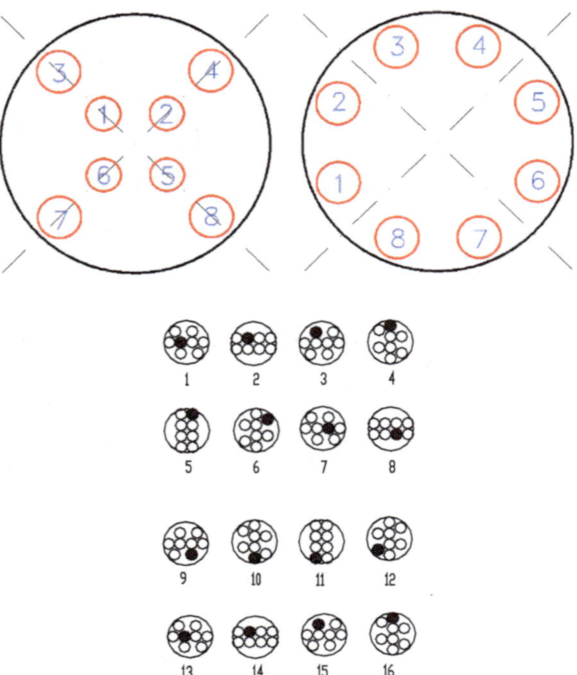


**Fig. 6.15** The cross-section of plait (8 strand) assemblies

The strands forming the braid are paired in mirror type; those migrating left or 'S' have a complement migrating right or 'Z'. The constitution of the S strands is identical to those Z strands except they are twisted in the opposite sense. The effect is to form a torque balanced structure, balanced at all loads.

Eight-strand plaited ropes are formed from four pairs of strands, the pair being constituted successively of two strands twisted in the S direction and then two strands twisted in the Z direction; they do not yield a void in the core since during the path alternating pairs do lie alongside each other at the centre of the rope.

Figure 6.16 shows the load extension of eight strand rope; also shown are the strands, yarns and fibres. This exhibits a smaller break strain and large break load.

## *6.3.1 The Geometry of Braids and Plaits*

The analysis of such structures is approximate since the structure is not geometry preserving in its detail; one can assume that over a pick, the distance along the rope between successive contacts on the yarn, the geometry is defined but not uniform whereas at each pick point and indeed at other pick spaced locations the geometry is repeated.

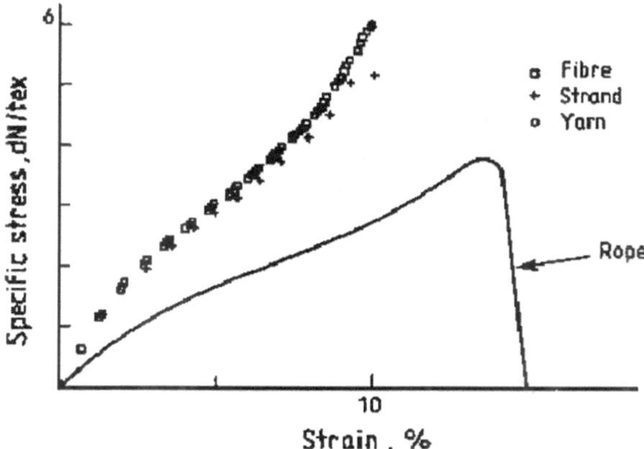


**Fig. 6.16** The extension of a 8 strand (plait) nylon rope

Over many pick stations it is observed that the profile described by the yarns is periodic since the geometry repeats at each station. An assumption that conforms to this observation is that the yarns follow a *sinusoidal* path, Fig. 6.13; this path assumption is not unique since there are other functions that are periodic and not harmonic.

Using this harmonic or sinusoidal assumption, the yarn path can be written as follows,

$$r = \frac{r_{max} + r_{min}}{2} + \frac{r_{max} - r_{min}}{2} \sin\frac{n\psi}{2} = r_{mean} + \frac{\Delta r}{2} \sin\frac{n\psi}{2}$$

where $r_{max}$, $r_{min}$ and $r_{mean}$ are the maximum, minimum and mean radial stations and $\Delta r$ is the radial travel of the yarns and $\psi$ is the angular station, shown as sweeping anticlockwise for the S yarns; $n$ is the number of S/L (or Z/R) yarns in the braid and there are a total of $2n$ yarns in the braid. The direction cosine formed by the yarn and the rope axis, is $\cos\theta$ $(= dz/ds)$ where $\frac{ds}{dz} = \sqrt{1 + \frac{dx^2}{dz} + \frac{dy^2}{dz}} = \sqrt{1 + \frac{d\psi^2}{dz}\left(r^2 \frac{dr^2}{d\psi}\right)}$.

Thus

$$r = r_{mean} + \frac{\Delta r}{2} \sin\frac{n\psi}{2} \quad \text{and} \quad \frac{dr}{d\psi} = \frac{n\Delta r}{4} \cos\frac{n\psi}{2}$$

At this point a deduction that the direction cosine is constant through the braid for each yarn is made; for a uniform portion of the braid, the geometry, load and deformation is the same at each repeat station, Fig. 6.9. There is thus no tendency to slip over the picks and consequently the load or tension carried by each yarn is constant through the braid; also the axial load carried by the assembly of yarns is constant at all stations. Thus it follows that the contribution from each yarn to the structure angle made by the yarn to structure load is the same at all stations and since the yarn load is constant it follows that the direction cosine is constant. At some parts the yarns are plunging to or rising from the rope axis and hence the cant angle between the yarns and the axis is reduced so that the total angle is constant; at those stations where one yarns crosses another, the plunge angle is zero and the cant axis is maximum.

There is an interesting consequence to this, leading to an equivalent braid radius; since the direction angle of a braiding yarn is constant through is path, it could be considered to behave as a geometry preserving helix with an equivalent radius $r_e$.

The analysis of braided constructions proceeds as follows: since the direction cosine is assumed constant then tan $\theta$ is a constant.

$$\text{Since } \tan\vartheta = \frac{d\psi}{dz}\sqrt{r^2 + \frac{dr}{d\psi}^2} \quad \text{then} \quad \tan\theta\int_0^L dz = \int_0^{2\pi}\sqrt{r^2 + \frac{dr}{d\psi}^2}\,d\psi$$

where $L$ is the pitch of the braid; for a specific number of braid yarns $n$, and for a defined radial geometry, $r_{mean}$ and $\Delta r$, the integrand can be quantified usually by numerical quadrature; there then results a reciprocal relation between the pitch length $L$ and the yarn angle $\theta$. Now defining a parameter $\lambda$ by $\Delta r/(2r_{mean})$ enables the above equation to be written as follows

$$\frac{L\tan\theta}{r_{mean}} = \int_0^{2\pi}\sqrt{\left(1 + \lambda\sin\frac{n\psi}{2}\right)^2 + \left(\frac{n\lambda}{2}\cos\frac{n\psi}{2}\right)^2}\,d\psi$$

Denoting the integral on the right hand side of the above by $I(\lambda,n)$, then the equivalent radius $r_e$ is as follows, $r_e = \frac{L\tan\theta}{2\pi} = r_{mean}\frac{I(\lambda,n)}{2\pi}$.

For those configurations in which the radial geometry is defined the above equation yields the equivalent radius $r_e$ and the direction cosine, cos$\theta$. For example an eight (transversely stiff) strand plait construction ($n = 4$) consists of four pairs of strands, each pair working as one component and following its own path. Then $r_{mean} = \Delta r = d$, the strand diameter and $\lambda = 0.5$ and the integral $I(\lambda,n)$, can be numerically evaluated.

For braid constructions where a braid is formed around a core, the inner radius, $r_{min}$ is specified as the core outside radius and $\Delta r$ is $d$, $r_{mean} = r_{min} + d/2$ and $\lambda = d/(2r_{mean})$.

For transversely soft strands, their compaction into a braid circumferentially squashes these components, thickening them radially; here

$$r_{mean} = d\frac{2+\sqrt{2}}{8}\sqrt{\frac{n}{\cos\theta}} \quad \text{and} \quad \lambda = \frac{\sqrt{2}}{2+\sqrt{2}}$$

and

$$\cos^2\theta + \beta\cos\theta - 1 = 0 \;\; \text{where } \beta = \left(I(\lambda,n)npd\frac{2+\sqrt{2}}{8}\right)^2$$

yields the direction cosine and thus the equivalent radius $r_e$.

Finally for braid on braid layered constructions it has been found by numerical experimentation that the following iteration will yield the final geometry:

a) assume initially an average helix and/or the associated direction cosine;
b) from this the compaction can be determined, thus resulting in the radial geometry, $r_{mean}$ and $\Delta r$;
c) from these the integrand $I(\lambda,n)$ can be estimated;
d) using the pitch length $L$, the direction cosine, $\cos\theta$ can be determined;
e) this value can now be introduced into step a) above and the iteration can be continued until $\cos\theta$ from step d) is sufficiently close to that introduced in step a).

## 6.4 Bending of Helical Structures

The bending of a linear helical structure is a relatively unexplored problem; in this section two extreme assumptions are made, namely that there is no slip between components and secondly that there is no friction and thus completely free slipping. These two assumptions would form the limits of bending, the realistic behaviour falling somewhere between these two limits [10, 11].

### *6.4.1 General Kinematics Associated with Bent Structures*

As the shaft is bent the component is forced towards the zero friction path. Either the friction is very large and the component will not slip or the friction is assumed zero, resulting in a minimum length path, or geodesic.. In the former case there will be a variation of strain with angular station, minimum at the inside of the bend and maximum at the outside. The true path and associated strain distribution will be somewhere between these extremes. The two theories are examined in the following sections, but first the path kinematics must be established.

Define $\varphi$ as a polar angle that sweeps around the circumference and is to be considered as the independent coordinate; $r$ is the radius of the helix, $R$ is the radius of curvature to which the helix is bent, measured to the helix central axis, Fig. 6.17.

Let $L_o$ be the helix length when bent; then the connection between curvature and pitch is $\chi = (s/L_0)(L_0/R)$ where s is a coordinate that runs along the bent helix axis. The Cartesian coordinates of a point on the component of the bent helix are

$$x = (R + r\sin\phi)\sin\chi,\quad y = (R + r\sin\phi)\cos\chi \quad \text{and} \quad z = r\cos\phi.$$

Now let $\chi$ be written as a function of the independent coordinate $\varphi$; that is $s = L_o\ \varphi/(2\pi)$, $\varphi = 2\pi$ and $\chi = (L_o/R)\ \varphi/(2\pi)$.

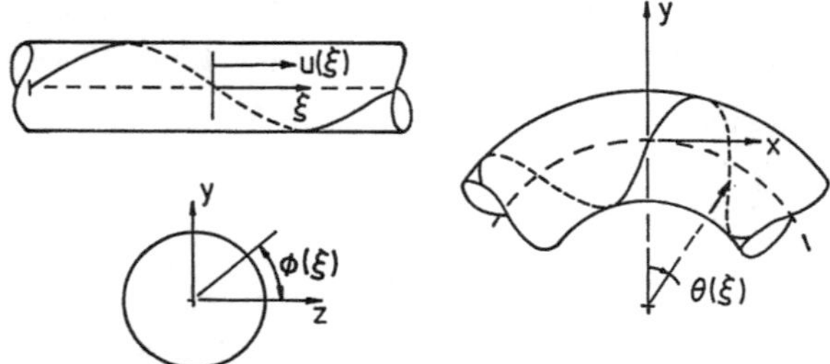


**Fig. 6.17** The bent helix

To find a line increment the following are required

$$\frac{dx}{d\phi} = (R + r\sin\phi)\cos\chi\frac{d\chi}{d\phi} + r\cos\phi\sin\chi,\ \frac{dy}{d\phi} = -(R + r\sin\phi)\sin\chi\frac{d\chi}{d\phi} + r\cos\phi\cos\chi$$

and $\frac{dz}{d\phi} = -r\ \sin\phi$.

The incremental line length is then $dl = \sqrt{\frac{dx}{d\phi}^2 + \frac{dy}{d\phi}^2 + \frac{dz}{d\phi}^2} d\phi = \sqrt{(R + r\sin\phi)^2 \frac{d\chi}{d\phi}^2 + r^2 d\phi}$.

## 6.4.2 No Slip (Friction) Bending

In this section the basic equations for axial structural force and bending moment for no slip bending are established.

As shown in the previous section the incremental line length $dl$ is given by

$$dl = \sqrt{(R + r\sin\phi)^2 \frac{d\chi}{d\phi}^2 + r^2 d\phi}$$

However the sweep angle $\chi$ is proportional to the helix independent coordinate $\varphi$; that is $s = R\chi$ and $\phi = \frac{2\pi s}{L} = \frac{2\pi R\chi}{L}$ so that $dl = \sqrt{(R + r\sin\phi)^2 \left(\frac{L}{2\pi R}\right)^2 + r^2 d\phi}$.

The initial incremental line length $dl_o$ is $dl_o = \sqrt{\left(\frac{L}{2\pi}\right)^2 + r^2 d\phi}$ and the initial direction cosine $\cos\ \theta_o$ is given in the following $\frac{1}{\cos\theta_o} = \frac{dl}{dL} = \sqrt{1 + (2\pi pr)^2}$ where $L$ is the pitch length and $p$ is the pitch rate (turns/unit length); writing $q = L/(2\pi)$, the component strain can be evaluated, $\varepsilon = \frac{dl}{dl_o} - 1 = \frac{\sqrt{(1+\lambda\sin\phi)^2 + \left(\frac{r}{q}\right)^2}}{\sqrt{1+\left(\frac{r}{q}\right)^2}} - 1$, and the direction cosine being, $\cos\theta = \frac{1}{\sqrt{(1+\lambda\sin\phi)^2 + \left(\frac{r}{q}\right)^2}}$.

The axial force $dF$ and the bending moment $dM_a$ about the neutral axis (the helix axis) is thus $dF = \sigma(\varepsilon)\cos\theta dA$ and $dM_a = \sigma(\varepsilon)\cos\theta r\sin\phi dA$.

If there are many small components used to form the structure then the incremental area $dA$ becomes $d\ r\ d\varphi$ where $d$ is the diameter of the component; these integrations can be evaluated,

$$F = \int_0^{2\pi}(\sigma rd)\cos\theta d\phi \text{ and } M_a = \int_0^{2\pi}(\sigma r^2 d)\sin\phi\cos\theta\, d\phi$$

where both $\sigma$ and $\cos\theta$ are functions of the angle $\varphi$; these integrals can be evaluated since for elastic behaviour of the components, $\sigma = F(\varepsilon)$.

### 6.4.3 Geodesic (No Friction) Bending

Since it is assumed that there is no friction, the component cannot carry a varying tension and hence the strain within the component is constant (independent of $\varphi$).

The length of this component in a pitch length $L_o$ is

$$l = \int_0^{2\pi}\sqrt{(R + r\sin\phi)^2\frac{d\chi}{d\phi}^2 + r^2}d\phi = \int_0^{2\pi} I(\phi, \frac{d\chi}{d\phi})d\phi.$$

For a minimum length or a geodesic path the following is necessary,

$$\frac{d}{d\phi}\frac{\partial I}{\partial\left(\frac{d\chi}{d\phi}\right)} - \frac{\partial I}{\partial\chi} = 0.$$

Since **MI/M**$\chi = 0$, if follows that

$$\frac{\partial I}{\partial\left(\frac{d\chi}{d\phi}\right)} = \frac{(R + r\ \sin\phi)^2\frac{d\chi}{d\phi}}{\sqrt{(R + r\ \sin\phi)^2\frac{d\chi}{d\phi}^2 + r^2}} = A$$

where $A$ is a constant (of integration); solving for $d\chi/d\varphi$ yields the following

$$\frac{d\chi}{d\phi} = \frac{Ar}{(R + r\ \sin\phi)\sqrt{(R + r\ \sin\phi)^2 - A^2}} = \frac{B\lambda}{(1 + \lambda\ \sin\phi)\sqrt{(1 + \lambda\ \sin\phi)^2 - B^2}}$$

where $\lambda = r/R$ and $B = A/R$; the pitch length $L = R(\chi_\varphi = 2\pi\text{-}\chi_\varphi = 0)$ is then given by the following

$$\frac{L}{r} = \int_0^{2\pi}\frac{B}{(1 + \lambda\ \sin\phi)\sqrt{(1 + \lambda\ \sin\phi)^2 - B^2}}d\phi$$

and the component length l for the helix length is then

$$l = \int_0^{2\pi} \frac{dl}{d\phi} d\phi = R \int_0^{2\pi} \sqrt{(1 + \lambda \sin\phi)^2 \frac{d\chi^2}{d\phi} + \lambda^2 d\phi}.$$

Substituting for $d\chi/d\varphi$ gives the final equation

$$\frac{l}{r} = \int_0^{2\pi} \frac{1 + \lambda \sin\phi}{\sqrt{(1 + \lambda \sin\phi)^2 - B^2}} d\phi.$$

#### 6.4.3.1 The Significance of the Constant *B*

Recalling that

$$\frac{d\chi}{d\phi} = \frac{B\lambda}{(1 + \lambda \sin\phi)\sqrt{(1 + \lambda \sin\phi)^2 - B^2}} = \frac{B\lambda}{\sqrt{1 - B^2}} \, for\, \phi = 0$$

And that $d\chi = ds/R = \lambda\, ds/r$ and $dc = r\, d\varphi$ where $dc$ is an increment in the circumferential direction, then the direction $\theta$ of the component at $\varphi = 0$ is given by the following

$$\cot\theta = \frac{ds}{dc} = \frac{1}{\lambda}\frac{d\chi}{d\phi} = \frac{B}{\sqrt{1 - B^2}}, \text{ and hence } \cos\theta_{\phi=0} = B$$

The scheme is formalized by scanning for a given $\lambda$, through a range of $B$'s in the range (0#*B*#1) to find that value that yields the corresponding value of *L*.

#### 6.4.3.2 The Direction Cosine cos*θ*

The direction cosine is given by $\cos\theta = dL/dl$ where $dL = R\, d\chi\, (1 + \lambda\, \sin\varphi)$, the incremental length parallel to the helix axis at the angular station $\varphi$. Substituting for the $dL$ and $dl$ in the above gives the following

$$\cos\theta = R(1 + \lambda \sin\phi)\frac{d\chi}{dl} = R(1 + \lambda \sin\phi)\frac{\frac{d\chi}{d\phi}}{\frac{dl}{d\phi}} = \frac{(1 + \lambda \sin\phi)\frac{d\chi}{d\phi}}{\sqrt{(1 + \lambda \sin\phi)^2 \frac{d\chi}{d\phi}^2 + \lambda^2}}.$$

Noting that

$$\frac{d\chi}{d\phi} = \frac{B\lambda}{(1 + \lambda \sin\phi)\sqrt{(1 + \lambda \sin\phi)^2 - B^2}}$$

and substituting in the above expression for cos $\theta$ gives

$$\cos\theta = \frac{\cos\theta_{\phi=0}}{(1 + \lambda\sin\phi)}$$

Since the angle $\varphi$ ranges between 0 and $2\pi$, there is an upper bound on $\cos\theta_{\varphi=0}$; that is $\cos\theta_{\phi=0} \leq 1 - \lambda$. Larger values of cos $\theta_{\varphi=0}$ will force a compaction or buckling of the component at the inside of the bend, $\varphi$ = $-\pi/2$. Figure 6.18 shows in plan view the variation of the slipped path from the no-slip path, the latter being shown as a dashed line.

The above expression for the direction cosine, cos $\theta$ suggests that it is largest at the inside of the bend and smallest at the outside. The development of the initially unbent helix is also shown and for a given station at the inside, the component slides forward on its travel from the bend inside to the outside of the bend, and conversely it slides back when travelling from the outside to the inside of the bend. Also since the direction cosine is inversely proportional to $(1 + \lambda\ \sin\varphi)$, on a development of the bent helix, the direction cosine is constant and the geodesic is a straight line.

#### 6.4.3.3 The Component Strain

Since the component length is given by

$$\frac{l}{r} = \int_0^{2\pi} \frac{1 + \lambda\ \sin\phi}{\sqrt{(1 + \lambda\ \sin\phi\,)^2 - B^2}}\, d\phi,$$

a simple numerical integration gives the component length, $l$ for a given curvature ratio $\lambda$, and for a given direction cosine at $\varphi = 0$; this direction cosine is selected to give the pitch length $L$ that corresponds to the unbent helix on the assumption that bending of the helix does not increase its' length.

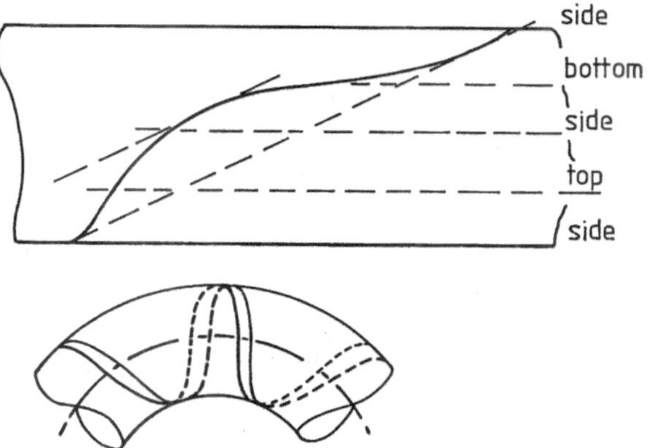


**Fig. 6.18** The bent helix, and the opened projection

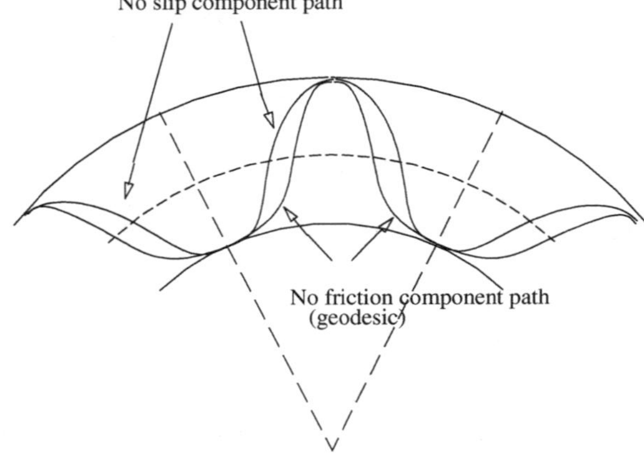


**Fig. 6.19** The limiting paths for component over a bent assembly

Figure 6.19 illustrates the zero friction and no slip paths for a component wound around a shaft (core). The results of the integration shows that the straight helix has maximum component length and that the result of bending the helix reduces the component length. Thus an interesting event can be used to explain the phenomenon, that a helical structure, when twisted to tighten the helix will tend to bend and strain relieve the components used to form the structure.

#### 6.4.3.4 Structure Forces and Moments Due to Component Strain

Since the strain in the component is constant along the component length, it is convenient to introduce the component stress $\sigma$ also constant. However since the component shortens when the structure is bent this stress becomes compressive with increasing structure curvature; the moment due to the component stress about the centre of curvature of the bent structure is

$$dM_c = \sigma \cos\theta(R + r\sin\phi)dA = \sigma\frac{\cos\theta_{\phi=0}}{1 + \lambda\sin\phi}(R + r\sin\phi)dA = \sigma\cos\theta_{\phi=0}RdA,$$

where $dA$ is the area of the component measured perpendicularly to the structure axis. It is observed that the contribution to structure bending moment is constant and independent of the circumferential station $\varphi$. The contribution to axial force is

$$dF = \sigma\cos\theta dA = \sigma\frac{\cos\theta_{\phi=0}}{1 + \lambda\sin\phi}dA.$$

If there are many small components used to form the structure then the incremental area $dA$ becomes $d\ r\ d\varphi$ where $d$ is the diameter of the component; these integrations can be evaluated

$$M_c = \int_0^{2\pi} (\sigma r d) \cos\theta (R + r \sin\phi) d\phi = \int_0^{2\pi} (\sigma r d) \frac{\cos\theta_{\phi=0}}{1 + \lambda \sin\phi} (R + r \sin\phi) d\phi = 2\pi \frac{\sigma r^2 d}{\lambda} \cos\theta_{\phi=0}$$

and

$$F = \int_0^{2\pi} (\sigma r d) \cos\theta d\phi = \int_0^{2\pi} (\sigma r d) \frac{\cos\theta_{\phi=0}}{1 + \lambda \sin\phi} d\phi = 2\pi (\sigma r d) \cos\theta_{\phi=0} \frac{1}{\sqrt{1 - \lambda^2}}.$$

The bending moment measured about the structure neutral axis, or the helix axis is then $M_a = M_c - RF = 2\pi \frac{\sigma r^2 d}{\lambda} cos\,\theta_{\phi=0} \left(1 - \frac{1}{\sqrt{1-\lambda^2}}\right)$.

# References


1. Paul W (1970) Structural mechanics of ropes. Review of synthetic fibre ropes, US Coast Guard Academy, Research and Development Project
2. Hsu P (1985) Tensile behavior of three strand twisted rope. Presented to fiber society, Boston
3. Toney MM (1986) On the mechanical behavior of twisted synthetic ropes. MSc thesis, MIT
4. Backer S, Hsu P (1985) Structural mechanics of plied twisted structures with partial frictional constraints. In: Kawabata S et al. (eds) Proceedings of the third Japan Australia joint symposium on objective measurement applications to product design and process control
5. Leech CM (1986) The assembly and modelling of synthetic ropes. SECTAM XIII, South Carolina, 1986 and 30th Anniversary Conference of Institute of mechanics, Beijing, China
6. Leech CM (1986) The assembly and modelling of synthetic ropes using microcomputers In: Schrefler, Lewis (eds) Microcomputers in engineering, Pineridge Press, Nebraska
7. Knapp RH (1986) Computer aided design of cables. Sea Technology, Jul 1986
8. Leech CM (1987) Theory and numerical methods for the modelling of synthetic ropes. Commun Appl Numer Meth 3:187–194
9. Leigh Phoenix S (1974) Analysis of the mechanical behavior of a Kevlar49 tubular braided sleeve/core electromechanical cable. Naval Underwater Systems Center
10. LeClair RA, Costello GA (1986) Axial, bending and torsional loading of a strand with friction. In: Proceedings of 5th OMAE, Tokyo, 1986 (published by ASME, New York, vol 3, pp 550–555)
11. Raoof M, Hobbs RE (1984) The bending of spiral strand and armoured cables close to terminations. In: Chung JS, Lunardi VJ (eds) proceedings of 3rd International Offshore Mechanics and Artic Engineering Symposium, vol 2, ASME, New York, pp 198–205 (Also published in ASME J Energy Resour Technol 106(3):349–355)

# Chapter 7
# Contact Force and Friction

**Abstract** This chapter introduces contact forces and pressures and its modelling; consequent to this is friction as occurs within the structure and between the constituent components. Various friction or more correctly energy dissipative mechanisms are identified and are grouped into two categories; those that arise because of relative motion between components are labelled INTER modes and those that arise because of the deformation of the component are INTRA modes and dilation and distortion are in the latter category. Because of the inhomogeneous nature of the fibre structures continuum theories do not necessarily lead to the best models. The quantification of dilation and distortion is achieved through the packing factor and the shape factor. Also included in this chapter is the modelling of structure life as limited by the continued abrasion between components. This uses the friction models and various wear criteria which must be measured by testing.

In any fibre structure where there is no bonding agency or resin, the result of any deformation must result in a slip of contiguous components; this is especially true in the loading of ropes [1]. Since there are many fibres and thus bearing surfaces, friction is an extremely important aspect of loading. The components are forced together since they are twisted or plaited and the action of loading will increase any bearing pressures. The tendency to slip occurs because of the finite transverse dimensions of the components. The actual slip distance accompanying relative fibre motion is small and can only be a fraction of the fibre diameter whereas the contact pressure is a function of the twist/plait geometry. The slip and the contact pressure combine to give the work done by friction in opposing a structure deformation. Whereas the local slip work may be small, the number of slipping locations will be large and the resulting friction reaction may be significant. However the more important aspect of friction occurs because of the repeated action of loading/unloading; some ropes may be in site for many years and although the working load may be a small percentage (5–20 %) of the break load, the number of load cycles can be as high as 10 million (1 cycle/min for 20 years).

C. M. Leech, *The Modelling and Analysis of the Mechanics of Ropes*,
Solid Mechanics and Its Applications 209, DOI: 10.1007/978-94-007-7841-2_7,

The energy lost through the accompanying friction hysteresis is thus substantial and is dissipated through the rope as material fatigue, heat and component wear and abrasion.

In this chapter the analysis of contact forces and pressures is described; Coulomb friction is the model used to develop the friction work. The determination of the contact pressures is achieved by again the Principle of Virtual Work, this time where the virtual displacement is radial. Virtual slip then is used in the manner of virtual displacement to give an expression for virtual work due to friction and from this the forces necessary to overcome friction are determined.

When a virgin rope is loaded and then unloaded, it will not return to its' virgin length; successive loadings also result in an increase of the load-free length. This set is the result of many internal actions in the rope structure from the decrimping of the fibres and polymer chains to the resettlement of the components within the rope. They cannot be attributed to friction since although friction acts opposite during unloading to that in loading, there can be no set as at zero loads there is no contact pressure and thus no friction. However there can be a transverse set due to lateral compactions of the rope strands and this will be considered later in this chapter; there are few experimental results to guide the modelling of rope set.

## 7.1 Contact Forces and Pressures

These are developed for transversely discrete geometries; recall that the strain is

$$\varepsilon_c = (1+\varepsilon_s)\sqrt{\left(\frac{1+q^2}{1+{q_0}^2}\right)} - 1, \text{ where } \quad q = \frac{2\pi pr}{1+\varepsilon_s} \quad \text{and} \quad q_0 = 2\pi p_0 r_0.$$

The component strain energy/unit volume is $U^*(\varepsilon_c)$, and the structure layer strain energy/unit structure length $U_s$ due to the component is $U_s = U^*(\varepsilon_c)\pi d^2/4$ where $\varepsilon_c$ is the strain of the components in that layer and depends on the helix parameters $p$ and $r$ and on the structural strain $\varepsilon_s$.

The Principle of Virtual Work has been applied earlier; the variation of the strain energy for a structure length $L_0$ was equated with the virtual work done by system strain and twist and was applied to give the external force system required to deform the helix and thus the structure, the force components in this instance being the axial load $F$ and the torque $T$.

Now a total contact force $p_{tc}$, is introduced so that its' action is to resist changes in helix radius of each component within a specific layer, Fig. 7.1. The virtual work done by this external action is $\delta W = p_{tc}\,\delta r$ where $\delta r$ is the virtual radial displacement and where the action of $p_{tc}$, is to resist the inward motion of each component towards the centre of the axis of the helix; it is a line force acting over the length of the helix or alternatively the length of the component.

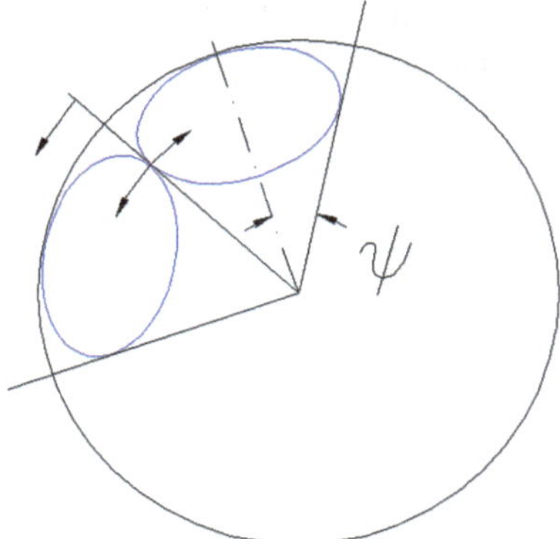


**Fig. 7.1** The contact between contiguous components

The Principle of Virtual Work when applied to a given structure length $L_0$ equates the virtual work with the variation of strain energy $\delta U$, as following

$$0 = \delta W - \delta U = P_{tc}\,\delta r - L_0\,\delta U_s$$

The variation of the structure strain energy density $\delta U_s$ has been expressed in terms of the virtual strain $\delta\varepsilon_s$, and this was in turn written in terms of the virtual stretch $\delta L$ and the virtual twist $\delta\varphi$; it now needs to be written in terms of $\delta r$ as follows

$$\delta U_s = \frac{\partial U_s^*}{\partial \varepsilon_c}\,\delta\,\varepsilon_c \quad \text{where} \quad \delta\,\varepsilon_c = \frac{\partial \varepsilon_c}{\partial \varepsilon_s}\,\delta\,\varepsilon_s + \frac{\partial \varepsilon_c}{\partial q}\,\delta q$$

and where $\delta\,\varepsilon_s = \dfrac{\delta L}{L_0}$ and $\delta q = 2\pi\left(\dfrac{p\delta r + r\delta p}{(1+\varepsilon_s)} - \dfrac{pr}{(1+\varepsilon_s)^2}\,\delta\,\varepsilon_s\right)$

The important term required is the coefficient of $\delta r$; defining the contact pressure $p_{0c}$ as the contact force/unit contact length, taken as the initial helix length $L_0$ leads to the following expression

$$P_{tc} = p_{0c}\,L_0 = \frac{L_0\,\pi\,d^2}{4}\frac{\partial\,U^*}{\partial\,\varepsilon_c}\frac{(1+\varepsilon_s)q}{\sqrt{((1+q_0^2)(1+q^2))}}\left(\frac{2\pi p}{1+\varepsilon_s}\right)$$

and simplifying for $p_{0c}$ gives

$$p_{0c} = \frac{\pi\,d^2}{4}\frac{\partial\,U^*}{\partial\,\varepsilon_c}\frac{r}{(1+\varepsilon_s)}\frac{[2\pi p]^2}{\sqrt{\left(1+[2\pi\,p_0\,r_0]^2\right)\left(1+\left[\frac{2\pi pr}{1+\varepsilon_s}\right]^2\right)}}$$

In the above the contact force (units being force/length) is multiplied by the initial helix length $L_0$ to give a force; this definition is not unique since the contact force could be normalised by

(a) the initial helix length $L_0$, and the contact pressure being $p_{0c}$, shown above,
(b) the current helix length $L$, when the contact pressure is $p_c$,

$$p_c = \frac{p_{0c}}{1+\varepsilon_s} = \frac{\pi\, d^2}{4} \frac{\partial\, U^*}{\partial\, \varepsilon_c} \frac{r}{(1+\varepsilon_s)^2} \frac{[2\pi p]^2}{\sqrt{\left(1+[2\pi\, p_0\, r_0]^2\right)\left(1+\left[\frac{2\pi pr}{1+\varepsilon_s}\right]^2\right)}}$$

(c) the initial component length $l_0$, and the contact pressure is $p_{l0c}$,

$$p_{l0c} = p_{0c} \cos\theta_0 = \frac{\pi\, d^2}{4} \frac{\partial\, U^*}{\partial\, \varepsilon_c} \frac{r}{1+\varepsilon_s} \frac{[2\pi p]^2}{\left(1+[2\pi\, p_0\, r_0]^2\right)\sqrt{\left(1+\left[\frac{2\pi pr}{1+\varepsilon_s}\right]^2\right)}}$$

(d) the current component length $l$, when the contact pressure $p_{lc}$,

$$p_{lc} = p_c \cos\theta = \frac{\pi\, d^2}{4} \frac{\partial\, U^*}{\partial\, \varepsilon_c} \frac{r}{(1+\varepsilon_s)^2} \frac{[2\pi p]^2}{\sqrt{1+[2\pi\, p_0\, r_0]^2}\left(1+\left[\frac{2\pi pr}{1+\varepsilon_s}\right]^2\right)}$$

The selection of normalising length depends on the application of the contact force and the direction of bearing; in the above, the contact pressure exerted by components in one layer on those in an inner layer is quantified.

If the structure is assembled from large components, the contact pressure acts circumferentially in the structure, acting between components within the same layer, Fig. 7.1 Recalling that the included angle for n components is $\Psi\,(=2\pi/n)$, then the tangential contact pressure $p_t$ is given by the following $p_t = \frac{p_c}{2\sin\Psi}$.

Since the hard components follow the larger diameter helix they will exert a larger bearing force on their neighbours.

## 7.2 Friction and Relative Slip: Inter Modes

The first category of friction in linear fibre structures arises from relative slippage between components within the structure and where there is a contact or bearing force normal to the slip surface; this is defined as *inter* mode friction since friction acts between two contiguous components at the same level in a branch of the hierarchical tree or between contiguous components at equivalent levels in different branches. It cannot act between components at different hierarchical levels in the same branch. Four Inter Modes are identified and considered. To develop the theory of friction within these linear structures, it is necessary to introduce and develop the underlying mechanics principle. Again the principle of virtual work is employed, this time including that (nonconservative) work done against friction; the virtual work done against virtual displacements is $\delta W = F\delta L + T\delta\phi$. The contact forces are not included here since they are equated to the internal contact pressures and these two together are self-equilibrating (and self-cancelling) [2, 3].

If friction is present then the virtual work done against friction, written as an internal energy dissipation $\delta W_f$ appears as follows

$$0 = \delta W - \delta W_f - \delta U = F\delta L + T\delta\phi - \delta W_f - L_0\, \delta U_s$$

It now remains to quantify $\delta W_f$ in terms of the relative displacement or slip and the generalised friction force and this is described separately for each of the four modes.

Friction is dependent on the contact force, $N$; it is also dependent on the relative slip velocity vector $\underline{v}$,

$$\underline{F} \equiv \underline{F}(\underline{N}, \underline{v}, \ldots.).$$

The velocity slip vector is included as most friction is speed dependent, but all friction is dependent on the velocity direction. Coulomb friction opposing the tendency to move can be written $\underline{F} = -\mu N\, direction(\underline{v})$, where; the friction coefficient is assumed constant. This could be more succinctly written as $F = -\mu N \tanh(\lambda v)$. where the vector notation is dropped for simplicity and friction acts opposite to the motion. The hyperbolic tangent is introduced as a model to yield the directionality of the motion and to remove the singularity in the vicinity of zero motion; this asymptotes to +1 for positive motion and to $-1$ for negative motion. The steepness of the function in the neighbourhood of zero motion is governed by the constant $\lambda$.

There are two aspects of friction; first the resistance to move and second the force due to motion. In the first, the force necessary for motion is $F = \mu N$. If the force applied is less than $\mu N$ there is no motion and if it is equal to $\mu N$ there is the onset of motion. The force cannot be larger than $\mu N$, Coulomb friction. The second aspect and this is important here is the force due to motion. In the analysis of fibre structures the structure deformation is prescribed and the forces necessary to deform the structure are determined. For friction considerations the motion or slip is prescribed and the friction force is estimated. Thus the friction force is limited to $\pm\, \mu N$, and the principal purpose of the modelling is to estimate the normal force $N$ and the quantitative effect of the resulting friction force on the structure mechanics.

### 7.2.1 Mode 1: Axial Slip

When a structure has several components in the same layer and in contact with each other, is stretched, there is a slip between these components acting parallel to the component axis but along the contact line. The application of contact force between the contiguous components and the relative slip results in friction to oppose this motion, Fig. 7.2. This friction acts axially but in opposite directions on

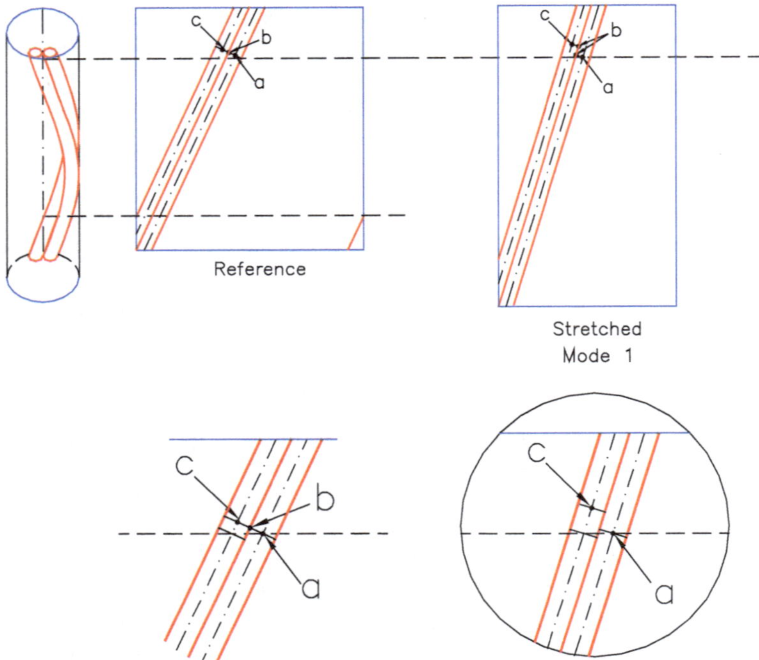


**Fig. 7.2** *Mode 1* Axial slip between components

opposite contact surfaces of the component, producing within each components a shear or distributed couple about the structure radius through the component. It is caused by stretching and/or twisting the structure.

Consider the structure development shown in Fig. 7.2 and the two neighbouring/contacting components indicated. The components have a finite diameter and it is this dimension that gives the slip a nonzero value; if the structure is stretched without twist, the relative separation of the location of the position on each component that were in contact at zero deformation. If the structure is also twisted the detail is similar, only canted to show the change in helix angle. In Fig. 7.2 the label b denotes a sample contact point and a and c are station lines on the contiguous components. This also shows the relative slip between these components.

If $d$ is the diameter of the component then the slip $S$ due to component strain $\varepsilon_c$ and change in component direction to $\theta$ from the initial attitude $\theta_0$ is $S = d([1 + \varepsilon_c]\tan\theta_0 - \tan\theta)$.The virtual slip $\delta S$ is thus $(\tan\theta_0\,\delta\,\varepsilon_c - \delta(\tan\theta))d$. To evaluate the virtual slip, and recalling $\tan\theta = dr/L$, then $\delta(\tan\theta) = (r/L)\delta\phi - (\phi r/L^2)\delta L$.

Also since $\delta\,\varepsilon_c = \frac{\partial\,\varepsilon_c}{\partial L}\,\delta L + \frac{\partial\,\varepsilon_c}{\partial\phi}\,\delta\phi$

then $\delta S = d\left[\left(\tan\theta_0\frac{\partial\,\varepsilon_c}{\partial L} + \frac{\phi r}{L^2}\right)\delta L + \left(\tan\theta_0\frac{\partial\,\varepsilon_c}{\partial\phi} - \frac{r}{L}\right)\delta\phi\right]$

Now recalling that

$$\tan\theta_0 = 2\pi p_0 r_0 = \frac{\phi_0 r_0}{L_0}$$

$$\text{and}\frac{\phi r}{L} = \frac{2\pi p r}{1+\varepsilon_s}$$

results in

$$\delta S = d\left(2\pi p_0 r_0 \frac{\partial\varepsilon_c}{\partial L} + \frac{2\pi p r}{L_0(1+\varepsilon_s)^2}\right)\delta L + d\left(2\pi p_0 r_0 \frac{\partial\varepsilon_c}{\partial\phi} - \frac{r}{L_0(1+\varepsilon_s)}\right)\delta\phi$$

where

$$\frac{\partial\varepsilon_c}{\partial L} = \frac{1}{L_0\sqrt{1+(2\pi p_0 r_0)^2}\sqrt{1+\left(\frac{2\pi p r}{1+\varepsilon_s}\right)^2}}$$

and

$$\frac{\partial\varepsilon_c}{\partial\varphi} = \frac{2\pi p r^2}{L_0\sqrt{1+(2\pi p_0 r_0)^2}\sqrt{1+\left(\frac{2\pi p r}{1+\varepsilon_s}\right)^2}}$$

where the subscript$_0$ denotes the reference or initial (undeformed) configuration. Then the virtual slip can be written as follows

$$\delta S = [S_L]\delta L + [S_\phi]\delta\phi$$

where $S_L$ and $S_\varphi$ are the mechanical gains in motion between slip and structure stretch and twist.

At this point an assumption as to the nature of friction must be made; the contact pressure $p_{l0c}$ has been quantified, this pressure being the component normal pressure based upon the initial component length $l_{0c}$. If Coulomb friction is assumed so that the friction force is given by a friction coefficient, $\mu_{AS}$ times the contact force, the virtual work expression becomes

$$\delta W_f = \mu_{AS} p_{l0c}\, l_0 (S_L\,\delta L + S_\phi\,\delta\phi)$$

The coefficient multipliers of $\delta L$ and $\delta\varphi$ are used to augment the expressions for end load and torque.

Figure 7.3 shows the effect of friction for transversely hard and soft components; the test configuration is again the three component structure, each component being a 1 mm diameter nylon fibre twisted with a pitch 100 tpm. The nylon is also assumed to have a breaking strain 0.25, and an elastic (linear) modulus 2 GN/m$^2$.

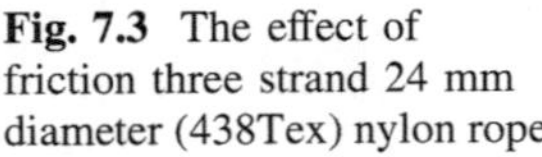
**Fig. 7.3** The effect of friction three strand 24 mm diameter (438Tex) nylon rope

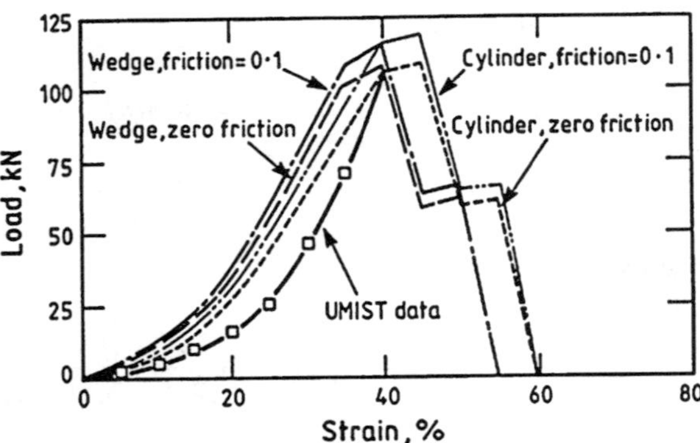


Figure 7.3 shows the effect of structure strain on the inter-component friction, the friction and load curves being additive. The friction coefficient has been assumed to be 1, and even with this large value the effect of friction on structure is negligible. The importance of friction arises in the cyclic loading of these structures and results in large (for large multi-component ropes) energy absorption and in component wear. Also shown, Fig. 7.4 is the plot of estimated contact forces for the same configuration.

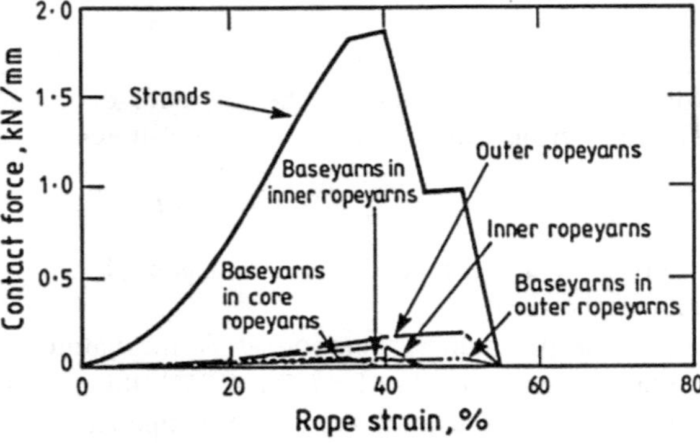


**Fig. 7.4** The contact forces for component types in the 24 mm three strand rope (strands, ropeyarns and baseyarns)

## *7.2.2 Mode 2: Component Twist*

Friction between components in the same layer but due to the tendency of components to twist about their axis; each component carries a torque and unless this torque is totally resisted by the component ends at the structure end, it must be resisted by a friction torque that opposes unwinding, Fig. 7.5.

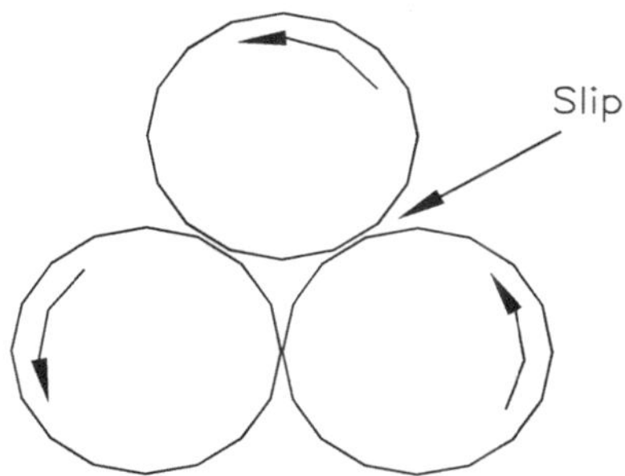


**Fig. 7.5** *Mode 2* Component twist

The degree of slip, in this case a twist is length dependent since the friction (reactive) torque developed is proportional to the component length. It is applicable to those structures assembled from incomplete component lengths, caused by either using short component lengths (staple fibres), a break in a component or join between components, or in the development of splicing. This mode is not generally useful for continuous, geometry preserving assemblies.

For geometry preserving structures the centre of rotation of each component is the centre of the structure and if each component rotates by the same amount there is no Mode 2 slip between components. If however the centres of rotation of the components are different to each other and to that of the structure there is a tangential slip between neighbouring components, Fig. 7.5. For a geometry preserving structure this mode is not present but if there is a break in or end to any component, this mode acts to prevent the structure unwinding. The action of unwinding is caused by the torque developed in the component and is resisted by friction on the slip surface. This mode of slip (and of friction) is applicable to splicing since there are strand ends that are fed through the splice and are held in position by contact pressure and friction.

Contact pressure has been considered previously; the equations for this are as follows, where the contact 'pressure' is a line force (units: N/m) and the length dimension depends on the measurement length. As before the two possibilities are the initial length (helix $L_0$, component $l_0$) and the current or stretch length ($L$, $l$).

(a) the initial component length $l_0$, and the contact pressure is $p_{l0c}$,

$$p_{l0c} = \frac{\pi d^2}{4}\frac{\partial U^*}{\partial \varepsilon_c}\frac{r}{1+\varepsilon_s}\frac{[2\pi p]^2}{\left(1+[2\pi p_0 r_0]^2\right)\sqrt{\left(1+\left[\frac{2\pi pr}{1+\varepsilon_s}\right]^2\right)}}$$

(b) the current component length $l$, when the contact pressure $p_{lc}$,

$$p_{lc} = \frac{\pi d^2}{4}\frac{\partial U^*}{\partial \varepsilon_c}\frac{r}{(1+\varepsilon_s)^2}\frac{[2\pi p]^2}{\sqrt{1+[2\pi p_0 r_0]^2}\left(1+\left[\frac{2\pi pr}{1+\varepsilon_s}\right]^2\right)}$$

The action of this contact pressure is towards the centre of the helix and would applied for Mode 2 slip against an inner component or layer; $p_{l0c}$ is the more useful as lengths are usually measured in the reference state.

For Mode 2 slip between components within the same layer, the contact pressure $p_t$ acts tangentially,

$$p_t = \frac{p_c}{2 \sin \Psi}$$

where $\Psi$ (= $2\pi/n$) is the included angle for $n$ components.

Recalling that the action of this friction is to resist the unwinding of a component, there is a given component torque that can remain in equilibrium with this friction,

$$\mathrm{T} = l_0 \, p_t \, \mu_{TW} \, d$$

where $T$ is the maximum or 'limiting' friction torque that a component can produce and can be retained by the structure; any larger torque developed by further twisting the structure would be shed by untwisting. The relevant friction coefficient, due to relative twist is $:_{TW}$. This torque is length dependent, thus requiring an end to initiate a length measurement.

### 7.2.3 Mode 3: Scissoring

This acts between components in different layers and between components in the same braided/plaited layers, Fig. 7.6. The slip in this case is angular, the relative change in component direction cosine between contiguous components; within the same layer of a braided/plaited construction the angle between those (half of the components) that are right laid or 'Z' and the others in the same layer that are left, 'S' laid is reduced when structure is stretched. Between layers, where a component from an outer layer is bearing on one in the inner layer, their respective directions will change under loading, causing an angular slip between these components. The action is local and the contact force, a point force has to be equated to the line contact force quantified above. The friction torque is again torque acting about an axis that goes through the contact point.

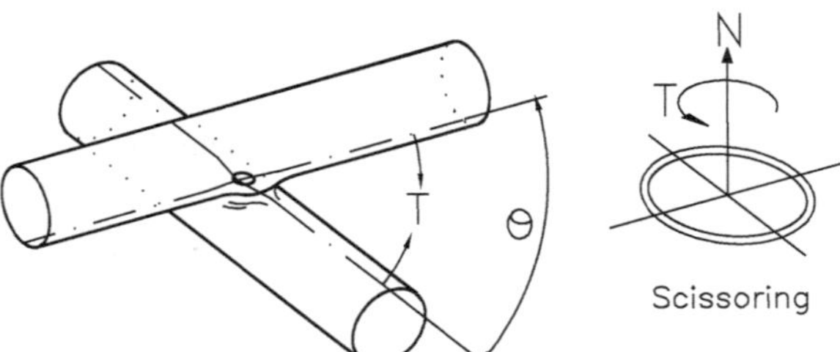


**Fig. 7.6** *Mode 3* Scissoring friction

Scissoring occurs when one component passes over another as occurring in braided and plaited constructions.

The angle between the two touching components changes (reduces) under increasing structure load in much the same way that the two blades of a scissor close. The contact pressure and the contact area increase with increasing load since the components are yielding and the contact is *advancing*.

To develop the effect of friction in resisting the change in scissor angle, consider the area of contact, Fig. 7.6; there is a normal pressure distribution $p$ that results from the normal force $N$ and the contact area $A$ as follows

$$N = \int_A p dA$$

The torque $T$ to overcome friction by scissoring is

$$T = \int_A \mu_c prdA$$

where $\mu_c$ is the Coulomb friction coefficient and $r$ is the location of the point of consideration from the torque axis. Introducing a non-dimensional pressure parameter, $f = \frac{\int_A prdA}{d \int_A pdA}$ then the torque can be written as $T = \mu_c fdN$ where $\mu_c$ and $f$ are dimensionless parameters; combining them to give a single dimensionless scissoring friction coefficient $\mu_{SC}(=\mu_c f)$ leads to the final friction relation

$$T = \mu_{SC} Nd$$

where $\mu_{SC}$ is associated with one set of component skewness, and surface softness and roughness (Fig. 7.7).

The previous development has assumed that the scissoring friction is independent of the direction of scissoring, or the opening and closing the scissors. The scissor angle $\theta$ defines the geometry of the component crossing, and thus for a given normal force $N$, the distance $z$ between the axes of the crossing component will vary with scissor angle. As the scissor angle opens, this distance will reduce and the approach increases. The approach rate, $dz/d\theta$ will decrease as the scissor

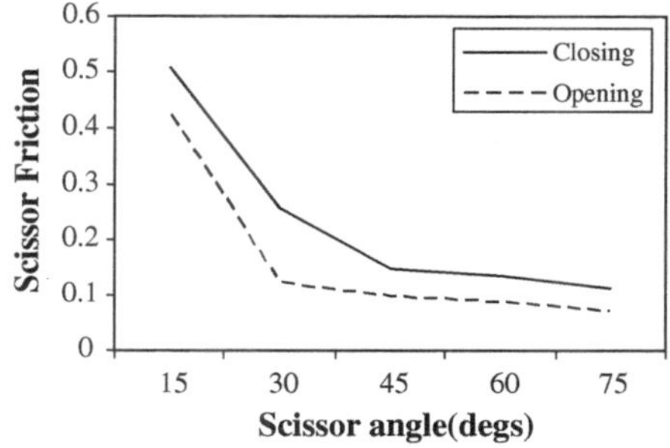


**Fig. 7.7** Scissoring friction for 1 mm polypropylene yarns

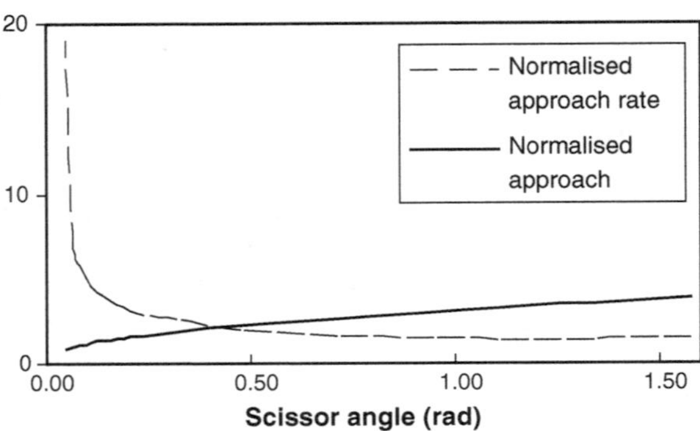


**Fig. 7.8** The compression of crossing cylinders

angle increases. These are shown in the following graph for homogeneous linear elastic cylinders; the approach distance $z$ is normalised by the material modulus, $E$, the cylinder diameter $d$ and the contact force $N$ such that the normalised approach is (zdE/N) (Fig. 7.8).

The virtual work done by a virtual displacement $\delta\theta$ in scissoring is $\delta W = T\delta\theta + N\delta z = \mu_{SC} N d\delta\theta$ or $\delta T\delta\theta = \mu_{SC} N d\delta\theta - N\delta z = N\left(\mu_{SC}\, d - {dz}/{d\theta}\right)\delta\theta$ thus yielding the following expression for the opening torque,

$$T = \left(\mu_{SC}\, d - \frac{dz}{d\theta}\right)N.$$

Note that the effect of the approach rate $dz/d\theta$ is negative since $z$ decreases as the angle increases or as the scissors open. This is usually determined experimentally although the above theoretical model could be employed as a gauge of model behaviour. For decreasing angles, scissor closing the friction acts in the opposite sense and the closing torque is

$$T = \left(\mu_{SC}\, d + \frac{dz}{d\vartheta}\right)N.$$

In both cases the torque T opposes the motion unless $dz/d\theta$ is sufficiently large and the scissors will open solely under the action of the normal force.

First consider the virtual displacement; recalling the helix angle $\theta$, where

$$\cos\theta = \frac{1}{\sqrt{1 + \left(\frac{2\pi pr}{1+\varepsilon_s}\right)^2}}$$

and by taking the variation of this with respect to the structure length $L$ gives

$$-\sin(\theta)\frac{\partial\theta}{\partial L} = \frac{1}{L_0}\frac{d}{d\,\varepsilon_s}\cos(\theta)$$

and hence

$$\frac{\partial\theta}{\partial L} = -\frac{2\pi pr}{L_0(1+\varepsilon_s)^2\left[1+\left(\frac{2\pi pr}{1+\varepsilon_s}\right)^2\right]}.$$

The virtual work is then

$$\delta W = T\frac{\partial\theta}{\partial L}\delta L = \left(\mu_{SC}\, d \mp \frac{dz}{d\vartheta}\right)N\frac{\partial\theta}{\partial L}\delta L$$

and the coefficient multiplier of $\delta L$ is the contribution of scissoring torque to the structure end load; it remains to give an expression for the normal contact force $N$ and to estimate the number of contacts in a pitch length $L_0$.

The contact force $N$ between the right and left helix components is estimated using the expression previously developed for radial contact force/unit contact length $p_{0c}$, as follows

$$p_{0c} = \frac{\pi\, d^2}{4}\frac{\partial\, U^*}{\partial\,\varepsilon_c}\frac{r}{(1+\varepsilon_s)}\frac{[2\pi p]^2}{\sqrt{\left(1+[2\pi\, p_0\, r_0]^2\right)\left(1+\left[\frac{2\pi pr}{1+\varepsilon_s}\right]^2\right)}}$$

The effective area helix length, the pitch over which this pressure acts at each crossover is $L_0/n$ where there are $2n$ components in the braid/plait, $n$ migrating left and $n$ right. There are $n^2$ crossings in the structure pitch and each component is involved in $n$ crossings. The contact force at each crossing is then

$$N = \frac{\pi\, d^2}{4}\frac{\partial\, U^*}{\partial\,\varepsilon_c}\frac{L_0\, r}{n(1+\varepsilon_s)}\frac{[2\pi p]^2}{\sqrt{\left(1+[2\pi\, p_0\, r_0]^2\right)\left(1+\left[\frac{2\pi pr}{1+\varepsilon_s}\right]^2\right)}}$$

where there $n$ such contacts per component in a pitch length $L_0$.

### 7.2.4 Mode 4: Sawing

Sawing acts between components in different layers and within braids and plaits although not in geometry preserving deformations. Sawing will occur in local structure deformation, specifically bending. Again the action is local and again the point contact force has to be estimated from the line contact force (Fig. 7.9).

Sawing occurs when one component slides over another in a direction that is significantly tangential; the contact is local and the same expression for contact pressure and normal force $N$ as used in Mode 3 is applicable as follows

**Fig. 7.9** *Mode 4* Sawing

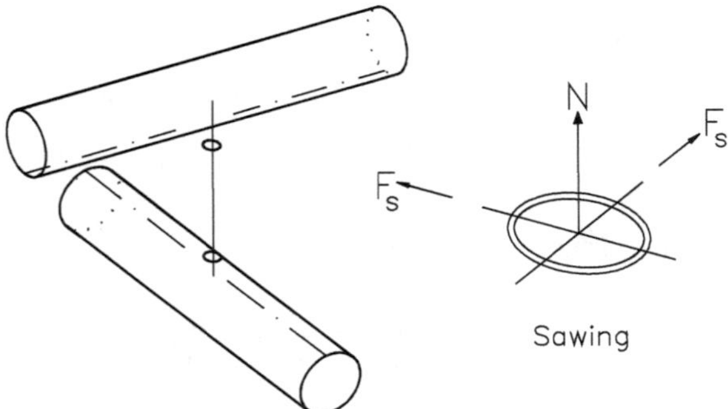


$$N = \frac{\pi\, d^2}{4}\frac{\partial\, U^*}{\partial\, \varepsilon_c}\frac{L_0\, r}{2n(1+\varepsilon_s)}\frac{[2\pi p\,]^2}{\sqrt{\left(1+[2\pi\, p_0\, r_0\,]^2\right)\left(1+\left[\frac{2\pi pr}{1+\varepsilon_s}\right]^2\right)}}$$

Sawing is connected with axial slip friction, Mode 1. This is because in sawing, only one of the two components can be sawed; the other experiences axial slip. If the angle between the two crossing components is $\theta$ along and one component is pulled over the other, the first component experiences axial slip friction. The direction of the friction force, opposing the motion is along this component in the direction of the motion. On the stationary component the same friction force is resolved along and across the component axis. That component along is again axial slip whereas across is sawing. Thus it is only possible to have pure sawing if the inclination between the two components is right angled and even then the sawing is only on the stationary component.

The friction sawing force $F_{SW}$ is as follows

$$F_{SW} = \mu_{SW}\, N$$

where $\mu_{SW}$ is the Coulomb friction coefficient; this force when combined with the virtual displacement $\delta x_s$ gives the virtual work done in sawing by each component on the other. The geometry of deformation is used to give the following

$$\delta\, x_s = \frac{\partial\, x_s}{\partial L}\,\delta L + \frac{\partial\, x_s}{\partial \phi}\,\delta \phi$$

and the force and torque required to cause friction slip is then

$$P = \mu_c\, N \frac{\partial\, x_c}{\partial L}$$

$$T = \mu_c\, N \frac{\partial\, x_c}{\partial \phi}$$

In order to estimate the above force $F$ and torque $T$, the partial derivatives must be evaluated and these can only be done if $x_c$ is known; for geometry preserving deformations these is no Mode 4 slip but this occurs for structure bending and especially for bending and flexure of braided and plaited structures. The detail of $x_s$ is thus specific to the structure configuration and deformation.

### 7.2.5 Transition Between the Sliding Modes

There is a geometric and a motion connection for Coulomb friction, between the sawing friction coefficient $\mu_{SW}$, Mode 4, the axial slip friction coefficient $\mu_{AS}$ Mode 1 and the component twist coefficient $\mu_{TW}$, Mode 2. To establish the connection, first consider the geometry. The crossing of the components is defined by the angle $\theta$, the included angle defining the attitude of the second component relative to the first component. The line of slip, the relative motion of the second component relative to the first is at an angle $\psi$. The slip motion for the Modes 1, 2, and 4 can be summarised by these two angles in the following table

To establish a relation between the Coulomb friction $\mu$ and the orientation ($\theta$) of the components and their relative motion ($\psi$), various material properties need to

| | Mode | $\theta$ | $\psi$ |
|---|---|---|---|
| Axial slip, $\mu_{AS}$ | 1 | 0 | 0 |
| Component twist, $\mu_{TW}$ | 2 | 0 | $\pi/2$ |
| Sawing, $\mu_{SW}$ | 4 | $\pi/2$ | 0 |
| | | $\pi/2$ | $\pi/2$ |

be measured. At this stage only the three friction coefficient, $\mu_{AS}$, $\mu_{TW}$ and $\mu_{SW}$ have been quantified and any competing function will use these data. Also since the arguments $\theta$ and $\psi$ are periodic in $\pi$ then functional approximation must also be periodic suggesting in the first instance that the sine and cosine functions be used as basis functions. If this data were sufficiently numerous, then a Fourier approximation (minimum error square fit) could be developed; however since there are only 4 data, a collocation fit must be established. Following the above argument, the minimum function

$$\mu = \mu_{SW} + \left(\frac{\mu_{AS} + \mu_{TW}}{2} - \mu_{SW}\right) \cos\theta + \left(\frac{\mu_{AS} - \mu_{TW}}{2}\right) \cos(2\psi - \vartheta)$$

is suggested as an initial fit for the Coulomb friction for contacting fibres, wires, rods etc., where $\theta$ is the angle between the components and $\psi$ is the direction of motion of one component relative to the other.

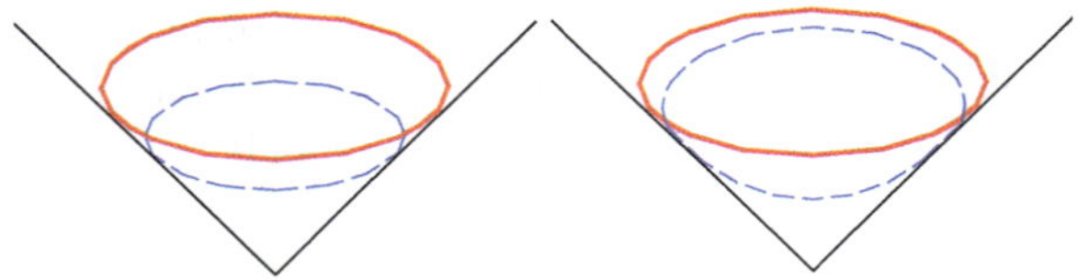

**Fig. 7.10** Component dilation $\Xi$, and distortion $\Omega$

## 7.3 Component Transverse Deformation and Set: Intra Modes

A second category of friction arises from the internal deformation of a component; for example if a strand is loaded so that it stretches or is transversely squashed and then unloaded, there is a hysteresis associated with the load cycle. This arises from the relative slips within the strand and whereas it could be related to the inter-slip between the subcomponents (in this case the yarns), this is not always quantifiable. This category of friction is called *intra* friction since it occurs within the components; it only arises from the motion of components in a higher hierarchical level in the same branch. Two *intra*-modes are identified and discussed, namely dilation and distortion (Fig. 7.10).

*Dilation*, $\Xi$ is the change in cross section area of the structure; when a yarn is stretched the cross section area is reduced by tightening the fibres against each other. There is also the material Poisson effect, the fibres radial strain being compressive when the axial strain is tensile, but this is only applicable at the lowest hierarchical level, the fibre level.

When the structure is loaded, the component is distorted in shape as well as in area. *Distortion*, $\Omega$ is this change in cross section shape (Fig. 7.11).

Nominally the cross section shape of these components is initially circular; when the structure is loaded these components press against each other with the result that there initial circularity is lost and the cross section shape progresses ultimately for soft structures to a wedge. The area shape change is due to localised transverse contact forces. When the structure is unloaded, the components will recover at least in part to their original state. They will not necessarily return to their initial area nor to their circular shape, the resulting hysteresis due to internal friction. Normally there is a lack of complete area recovery which results in a *set* or residual strain of the structure. This detail is shown in Sect. 6.2.1.

Figure 7.12 shows the effect of load on the reduction of cross section for an eight strand nylon rope. The rope strain is also shown as a comparative base.

The dilatation $\Xi$ and distortion $\Omega$ is determined by the structure deformation, but the effect of these two state variables on the structure mechanics is two fold. The first is geometrical and the second is mechanistic, invoking the energy of distortion and dilation.

There has been little if any research on the quantification of distortion and dilatation of yarn and strand assemblies.

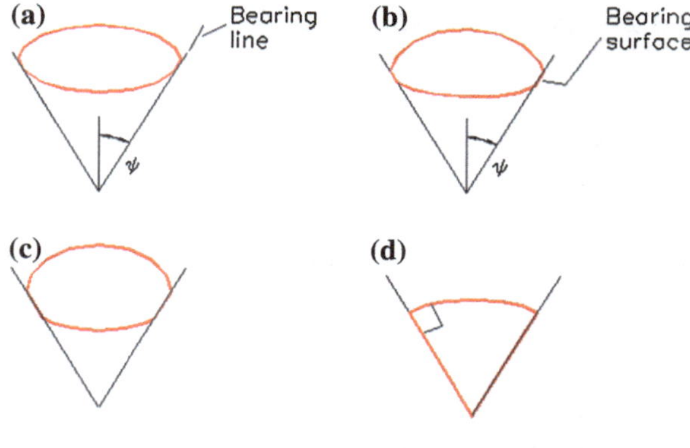


**Fig. 7.11** The progress of component distortion

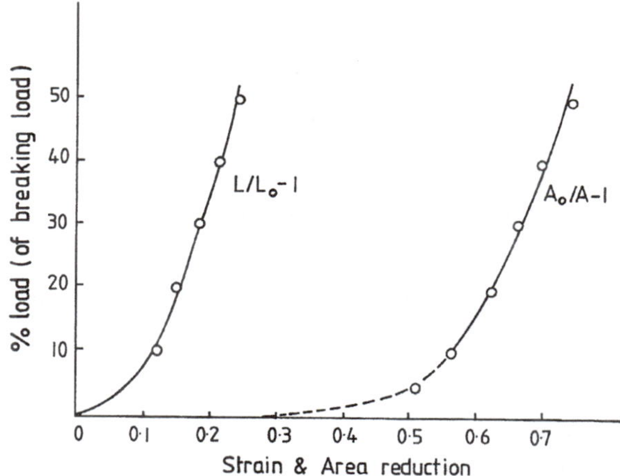


**Fig. 7.12** The reduction of area in an eight strand rope

First, a more pragmatic and more useful theory is introduced, and second, the classical approach is included for completeness and in outline since it is a possibility especially with the use of commercial finite element computer suites.

### *7.3.1 Dilation Measure, Ξ and Packing Factor $P_f$*

Recall that the definition of packing factor, $P_f$ is the ratio of component cross section area to the structure cross section area ($= A_c/A_s$). The packing factor is limited to between zero and one, the latter implying that there is no *'air space'* within the structure. There is direct correlation between dilation and packing factor and in the following the packing factor will be used to represent dilation.

### 7.3.2 Distortion Measure, Ω

To represent the distortion experienced by the compaction of a component due to structure loading, first recall that the component is initially circular. This is defined by an area $A\ (=\pi r^2)$ and a perimeter $P_e\ (=2\pi r)$. As the component is distorted, both these measures change; however since the change in cross section area is the dilation, the distortion can be measured solely by the change in perimeter. A distortion measure thus is constructed by the perimeter for a specified, constant area $(= P_e^2/4\pi A)$ This is minimum and equal to 1 for the circular cross section and increases when the cross section is non circular. To complete the establishment of the distortion measure, the *ln* function is introduced, $\Omega = Ln\left(P_e^2/4\pi A\right)$ such that $\Omega > 0$, and increases as the component deforms. For a wedge, included angle $\theta$, the maximum distortion is $Ln\left((2+\vartheta)^2\Big/2\pi\vartheta\right)$. For a three strand structure each strand will have a maximum distortion, $Ln\left((3+\pi)^2\Big/3\pi^2\right) = 0.2421$ and for more strands the maximum distortion will be larger.

### 7.3.3 The Set in Distortion and Dilation

The effect of distortion and dilation on the mechanics of the structure is geometrical, since for a given structure deformation, the component cross section will determine the component positioning, and thus the component stretch and twist. A highly deformed component with substantially reduced cross section area will sit closer to the axis, and thus will be subject to more deformation than a component that is further from the axis due to being less deformed.

Dilatation and distortion are likely to be history or path dependent and thus are not uniquely defined by the deformation state, especially when the structure is composed of transversely soft components. When loaded the components assume a compressed and distorted shape, and on unloading, the components cannot be assumed to return to their initial circular shape, the resulting hysteresis again due to internal friction.

#### 7.3.3.1 Dilation Set

Consider first the sequence of structure loading, in tension. The structure is stretched and this will result in component strain. The change in component length will result in a reduction in component diameter, and thus reduction in diameter and thus cross section area and is modelled by an increase in packing factor in the component. The relation between this packing factor and the stretch of the component, is typified by a functional relationship, $P_F = P[P_{FS}, \varepsilon, \ldots]$ where $P_{FS}$ is the initial packing factor, the packing factor before the onset of component loading and $\varepsilon$ is the component strain; there are other contributory measures that will also contribute to the drawdown, the reduction in cross section area of the component but those defined, the initial packing factor and the component strain are to most significant. It is a requirement that when the strain is zero the packing factor is the initial packing factor and for large strains the packing factor asymptotes to one.

One such model is shown, following, $P_F = P_{FS} + (1 - P_{FS})(1 - \exp(-D\varepsilon))$ where $D$ is a constant, specific to that component construction and constituent materials. At this point a parameter $\alpha$ is introduced, a measure of the progress of loading, such that $P_F = 1 - \exp(-\alpha)$ and $\alpha = -\ln(1 - P_F)$. A corresponding parameter $\alpha_0$, relating to the current packing factor $P_{FS}$, $\alpha_0 = -\ln(1 - P_{FS})$. The limits on the parameter $\alpha$ are $P_F = 0$, $\alpha = 0$, and $P_F = 1$, $\alpha = \infty$. A setting parameter for dilation $\lambda_\Xi$ is now introduced, such that if $\lambda_\Xi = 0$, there is no set, the unloading path is the same as the loading path, elastic dilation and if $\lambda_\Xi = 1$, the set is the same as the current packing factor, perfectly plastic dilation. The setting parameter is tied to another parameter, $\beta = 1 - (1 - P_{FS})\exp(\lambda_\Xi \alpha_0)$.

This will then establish the new transitional packing factor,

$$
\begin{aligned}
P_T &= \beta + (1 - \beta)(1 - \exp(-\lambda_\Xi \alpha)) = 1 - (1 - \beta)\exp(-\lambda_\Xi \alpha) \\
&= 1 - (1 - P_{FS})\exp(\lambda_\Xi[\alpha_0 - \alpha]).
\end{aligned}
$$

If the packing factor has increased, $P_T > P_{FS}$ then the new initial packing factor is updated to the transitional packing factor, $P_{FS} \leftarrow P_T$.

As the structure strain increases, the component is reduced in size as drawn into the available space and squashed; this is asymptotic, since there is a limitation in the amount of distortion and dilation that the component can experience. When the straining is stopped and the structure unloaded, the component will recover. The dilation will not necessarily follow in reverse, the dilation-straining path. It will in most cases follow a different path below the loading path, similar to that followed by the unloading of materials after exceeding the yield point. If there is complete unloading the unloading curve will cross the strain axis at a finite strain, the residual strain (Fig. 7.13).

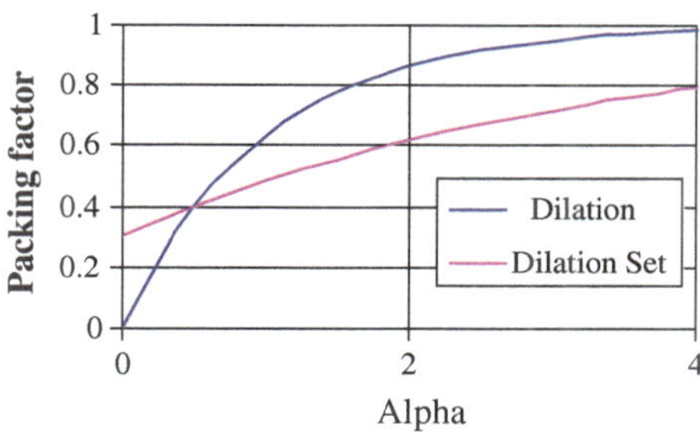


**Fig. 7.13** Dilation against structure loading

#### 7.3.3.2 Distortion Set

Again consider first the sequence of structure loading, in tension. The structure is stretched and this will result in the helix being stretched. The component follows the helix path and it will be both strained, due to induced load and bent. This load will try to pull the component into the space offered by its neighbours thus drawing it towards the structure axis. This will force it toward the wedge configuration thus causing component distortion. This distortion is thus caused by the curvature $\chi$ of the component path and by the component strain $\varepsilon$. If either of these parameters is zero then there will be to distortion. Thus the relation between distortion $\Omega$ and component curvature and the component stretch is typified by a functional relationship, $\Omega = \Omega[\Omega_S, \chi, \varepsilon, G\ldots.]$ where $\Omega_S$ is the initial distortion, that before the onset of component loading; G is a set of parameters that are geometrical important to the distortion process, and will include the effects of the number of components involved in the distortion process and the layering and braiding configurations. The requirement on this functional is that for zero strain and curvature and strain, the distortion is zero and for large strains and curvature the distortion asymptotes to its maximum, $\Omega_L$. One such model is shown, following,

$$\Omega = \Omega_S + (\Omega_L - \Omega_S)(1 - \exp(-C\chi\varepsilon))$$

$$\Omega = \Omega_S + (\Omega_L - \Omega_S)(1 - \exp(-C\chi(\varepsilon - \varepsilon_{\text{Res}})))$$

where $C$ is a constant, specific to that component construction, geometry and constituent materials. Again the dummy parameter $\alpha$ is introduced, a measure of the progress of loading, such that $\Omega = \Omega_L(1 - \exp(-\alpha))$ and $\alpha = -\ln(1 - \Omega/\Omega_L)$. The limits on the parameter $\alpha$ are $\Omega = 0$, $\alpha = 0$, and $\Omega = \Omega_L$, $\alpha = \infty$. A corresponding parameter $\alpha_0$, relating to the current distortion, $\Omega_S$, $\alpha_0 = -\ln(1 - \Omega_S/\Omega_L)$. The distortion setting parameter $\lambda_\Omega$ is now introduced, such that if $\lambda_\Omega = 0$, there is no set, and the unloading path is the same as the loading path, elastic distortion and if $\lambda_\Omega = 1$, the set is the same as the current distortion, perfectly plastic distortion. As for dilation, the setting parameter is tied to another parameter, $\beta$ as following

$$\Omega_S = \beta + (\Omega_L - \beta)(1 - \exp(-\lambda_\Omega\alpha)) = \Omega_L - (\Omega_L - \beta)\exp(-\lambda_\Omega\alpha)$$

Inverting gives $-\lambda_\Omega\alpha = \ln\left(\frac{\Omega_L - \Omega_S}{\Omega_L - \beta}\right)$ and $-\lambda_\Omega\alpha_0 = \ln\left(\frac{\Omega_L - \Omega_0}{\Omega_L - \beta}\right)$.; solving for $\beta$ gives $\beta = \Omega_L - (\Omega_L - \Omega_0)\exp(\lambda_\Omega\alpha_0)$. The limits on the distortion setting parameter $\lambda_\Omega$ are $\lambda_\Omega = 0$, $\beta = \Omega_0$, for elastic distortion and if $\lambda_\Omega = 1$, $\beta = 0$ since $\Omega_0 = \Omega_L(1 - \exp(-\alpha))$, perfectly plastic distortion.

This is then used to establish the new transitional distortion set, $\Omega_T$

$$\begin{aligned}\Omega_T &= \beta + (\Omega_L - \beta)(1 - \exp(-\lambda_\Omega\alpha)) = \Omega_L - (\Omega_L - \beta)\exp(-\lambda_\Omega\alpha) \\ &= \Omega_L - (\Omega_L - \Omega_0)\exp(\lambda_\Omega[\alpha_0 - \alpha])\end{aligned}$$

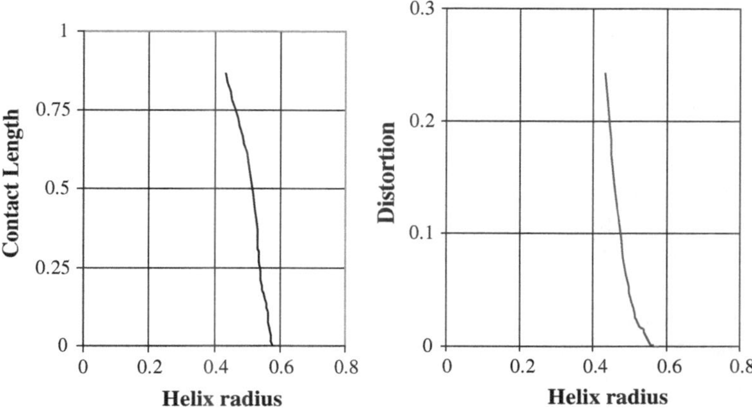


**Fig. 7.14** The distortion of a strand due to wedging

If the distortion has increased beyond the set distortion, $\Omega_T > \Omega_S$ then the new initial distortion set is updated to the transitional distortion, $\Omega_S \leftarrow \Omega_T$.

The Fig. 7.14 shows for a three stand assembly the change in contact length and the component distortion with the change in helix radius.

## 7.3.4 The Energy of Dilation and Distortion

The strain energy of deformation, specifically of dilation and of distortion, can be related back to the structure mechanics through the principle of virtual work. The strain energy of a component subject to stretch, twist, flexure, dilation and distortion can be written, for a *simple* component as

$$W_d\left(\varepsilon, \frac{d\theta}{ds}, \Xi, \Omega\right) = W_{d,e}(\varepsilon) + W_{d,t}\left(\frac{d\theta}{ds}\right) + W_\Xi(\Xi) + W_\Omega(\Omega)$$

Thus through the principle of virtual work,

$$\begin{aligned} \delta W &= P\delta L + T\delta\phi \\ &= \sum_{\text{All components}} \frac{\partial W_d}{\partial \varepsilon}\delta\varepsilon + \frac{\partial W_d}{\partial\left(\frac{d\vartheta}{ds}\right)}\delta\left(\frac{d\vartheta}{ds}\right) + \frac{\partial W_d}{\partial \Xi}\delta\Xi + \frac{\partial W_d}{\partial \Omega}\delta\Omega \end{aligned}$$

where the variations

$$\delta\Xi = \frac{\partial \Xi}{\partial \varepsilon}\delta\varepsilon + \frac{\partial \Xi}{\partial\left(\frac{d\vartheta}{ds}\right)}\delta\left(\frac{d\vartheta}{ds}\right) \text{ and}$$

$$\delta\Omega = \frac{\partial\Omega}{\partial\varepsilon}\delta\varepsilon + \frac{\partial\Omega}{\partial\left(\frac{d\vartheta}{ds}\right)}\delta\left(\frac{d\vartheta}{ds}\right)$$

require the quantities $\delta\Xi/\delta\varepsilon$, $\delta\Omega/\delta\varepsilon$, $\delta\Xi/\delta(d\theta/\mathrm{ds})$ and $\delta\Omega/\delta(d\theta/\mathrm{ds})$; this requires that the distortion and dilation be expressed in terms of the structure deformations. The energy of dilatation $U_\Xi$ and distortion $U_\Omega$ are functions of the dilatation and distortion. The derivatives $dU_\Xi/dL$, $dU_\Xi/d\varphi$, $dU_\Omega/dL$ and $dU_\Omega/d\varphi$ are then used to define the virtual work of dilation $\delta U_\Xi$ and of distortion $\delta U_\Omega$ in terms of the virtual displacements $\delta L$ and $\delta\varphi$. It is necessary that the energy of dilation and distortion be quantified, either by testing and measurements or by modelling and analysis. In a later section the foundations for analytical modelling of continuum components is broached, but this is very tentative; it needs the support of experimental measurements.

For polymer fibre structures, the geometric effect is very important as there is substantial transverse movement in these structures. For structures with more transversely stiff components, as used in wire rope, the first, the strain energy effect is more important.

### *7.3.5 The Continuum Model*

The following is a proposed theory for dilatation and distortion; as stated above there is no experimental validation for these proposed measures but they are derived from classical strain theories. The classical approach is to consider the structure as a continuum; clearly this is not a valid approach for a yarn composed of many yarns, but it could be useful for a structure where there is an infinite number of small components as considered in Chap. 4, [4].

First the following finite strains (in two dimensions, $r$ and $\theta$) are quoted,

$$\varepsilon_{rr} = \frac{\partial u_r}{\partial r} + \frac{1}{2}\left[\left(\frac{\partial u_r}{\partial r}\right)^2 + \left(\frac{\partial u_\theta}{\partial r}\right)^2\right]$$

$$\varepsilon_{\theta\theta} = \frac{1}{r}\frac{\partial u_\theta}{\partial\theta} + \frac{1}{2r^2}\left[\left(\frac{\partial u_r}{\partial\theta}\right)^2 + \left(\frac{\partial u_\theta}{\partial\theta}\right)^2 - 2u_\theta\frac{\partial u_r}{\partial\theta} + 2u_r\frac{\partial u_\theta}{\partial\theta} + {u_\theta}^2 + 2r u_r + {u_r}^2\right]$$

and

$$\varepsilon_{r\theta} = \frac{1}{2r}\left[\frac{\partial u_r}{\partial\theta} + r\frac{\partial u_\theta}{\partial r} - u_\theta + \frac{\partial u_r}{\partial\theta}\frac{\partial u_r}{\partial r} + \frac{\partial u_\theta}{\partial\theta}\frac{\partial u_\theta}{\partial r} + u_r\frac{\partial u_\theta}{\partial r} - u_\theta\frac{\partial u_r}{\partial r}\right]$$

The displacements $u_r$ and $u_\theta$ are functions of $r$ and $\theta$ and must be determined; one method of determining these is the employment of a finite element package. The functions that define $u_r$ and $u_\theta$ are either determined from experimental observations or estimated from component centre and boundary movements; then the finite strains can be quantified.

(a) The dilation mode: this occurs as a result of the change in cross section area in a component as it is stretched against the helix geometry and against the contiguous components.
The dilatation for an incremental area is

$$\Delta\Xi = (1 + \varepsilon_{rr})(1 + \varepsilon_{\theta\theta}) - 1$$

and the total area dilation $\Xi$ is

$$\Xi = \int_0^{2\pi}\int_0^{R(\theta)} (\varepsilon_{rr} + \varepsilon_{\theta\theta} + \varepsilon_{rr}\,\varepsilon_{\theta\theta}) r dr d\theta$$

(b) The distortion mode: this occurs as a result of the change in the shape of cross section area in a component.

Using the conventional theory associated with principal strains, the incremental distortion can be defined as follows

$$\Delta\Omega = \sqrt{(\varepsilon_{rr} - \varepsilon_{\theta\theta})^2 + (2\,\varepsilon_{r\theta})^2}$$

and the area distortion is

$$\Omega = 4\int_0^{2\pi}\int_0^{R(\theta)} \sqrt{(\varepsilon_{rr} - \varepsilon_{\theta\theta})^2 + (2\,\varepsilon_{r\theta})^2} r dr d\theta$$

## References

1. Howell HG, Mieszkis KW, Tabor D (1959) Friction in textiles. Butterworths Scientific Publications, London
2. Leech CM (2002) The modelling of friction in polymer fibre ropes. Int J Mech Sci 44(3):451–664
3. Johnson KL (1985) Contact mechanics. Cambridge University Press, Cambridge
4. Chung TJ (1988) Continuum mechanics (Chap. 2). Prentice-Hall International editions, NJ

# Chapter 8
# Component Wear, Life and Heating

**Abstract** In this chapter a preliminary investigation into component wear and structural heating is initiated, where the emphasis is on the wear and heating of the components due to repeated abrasion between contiguous components. This is considered by assuming that the wear and heating are related to the work done by sliding or scissoring. First the work done on the component by repeated sliding, sawing or scissoring is estimated and secondly the effect of this on the deterioration of component performance, wear is quantified. Finally the steady state thermodynamic equilibrium, due to repeated abrasion between contiguous components is determined.

In this section, the emphasis is on the wear and heating of the components due to repeated abrasion between contiguous components. This is considered by assuming that the wear and heating are related to the work done by sliding or scissoring. First the work done on the component by repeated sliding, sawing or scissoring is estimated and secondly the effect of this on the deterioration of component performance, wear is quantified. Finally the steady state thermodynamic equilibrium, due to repeated abrasion between contiguous components is determined.

## 8.1 The Work Done by Relative Motion Between Components

The action of relative motion against friction forces results in work done; generally the work done (Joules) is force (F) time distance (m) moved, shown in the following equation $\Delta W = F\Delta S$.

Recall that sliding friction when based on the Coulomb friction used the contact force N and the friction coefficient $\mu$; for the mode 1 friction, the contact force is a linear pressure force (units:- N/m) and for slip the slipping force is limited to μN (N/m). The longer the length the larger the pull force required but the contact force

C. M. Leech, *The Modelling and Analysis of the Mechanics of Ropes*, Solid Mechanics and Its Applications 209, DOI: 10.1007/978-94-007-7841-2_8,

is still limited to N and the work done to μN ΔS where ΔS is the relative movement between the components and the units of work done are Nm/m or J/m.

That is $\Delta W_{Sl} = \mu_{AS} N \Delta S$ and consequently the work done $\Delta W_S$ is the work done per metre of structure.

For sawing the contact zone between the components is identified by the orientation of one component with respect to the other and to the relative motion. If the relative motion is aligned to the orientation of one component then the contact is distributed along that component; the other component has a smaller fixed contact zone. Thus the friction effect is much more local and is expressed, $\Delta W_{SW} = \mu_{SW} N \Delta S$.

The work done here is associated with the respective contact zones, for each of contiguous components and could be quantified as work per unit area, J/m$^2$, or by the contact, that is J/node. As previously implied the effect of this abrasion could be quite different for each component. The components are subject to the same abrasion when the relative direction of motion of one component relative to the other bisects the orientation angle.

For the scissoring, the work done $W_{SC}$ is in opening or closing the scissors and is quantified by following equation $\Delta W_{SC} = T \Delta \theta$ where $\Delta \theta$ is the change in angle during that event. The friction torque $T$ opposes the scissoring motion at a node, that is the crossing of strands and consequently the work done here is per node, J/node.

## 8.2 The Friction Parameters

The work done on the components results in two primary consequences; the first is the wearing of the rope components at the friction areas, and the second is the heating of the components by repeated rubbing.

The estimation of work done is contingent on the friction law; the Coulomb friction law is assumed, and this requires as data the friction coefficient. The friction coefficient has to be measured and will depend on among other conditions,

(a) the rate of slip,
(b) the change of the surface due to wear and
(c) any lubrication or wetting of the surface.

It is assumed to be Coulomb friction, that is not to depend on the contact force, nor on the amount of slip. Obviously the contact force increases the contact area and in relatively transversely soft structures, rope strands, the large changes in contact area must affect the friction coefficient. Typical and very approximate values for friction coefficient are for linear sliding and sawing, $\mu_L = 0.1$ and for scissoring $\mu_S = 2$.

## 8.3 Wear

The surface abrasion of the contiguous strands occurs in three modes, sliding, scissoring and sawing. A wear criterion $W_C$ is proposed, the upper limit of work done which the component can accept without failing or requiring discard. The work done in wearing is the work done per event, $\Delta W$, times the number of events per cycle, $E$, times the number of cycles, $N$. The event is the single slip action in transition from one state or configuration to another.

That is for failure or discard, $NE\Delta W \geq W_C$ or the cycles to failure for a specific component is given as follows $N \leq W_C/E\Delta W$. This wear criterion must be measured; a wear criterion for polyester for failure in slip is typically and approximately taken as 500 $\mathrm{kJm^{-1}}$ and in scissoring, 300 kJ per node.

## 8.4 Steady State or Equilibrium Temperature, Surface Heat Convection

When a rope structure is subject to many load cycles, the internal abrasion sites generate heat. This heat, shown as temperature rises, will eventually exit the rope though the surface. At some state the heat generated by the internal friction points will balance the heat exiting and a steady state will exist. The internal friction points will then be responsible for raising the rope to that state and the surface conductivity is the mechanism by which the heat exits the system.

To develop this model, assume a cylinder model of a rope, diameter D with internal heat sources $q_i$ ($\mathrm{Jm^{-1}s^{-1}}$, $\mathrm{Wm^{-1}}$) distributed across a diametrical face. Also assume that the situation is axially independent, that is uniform along the cylinder length. All heat parameters are per unit length, per metre.

For a steady state condition, thermodynamic equilibrium, the heat input ($\mathrm{Jm^{-1}s^{-1}}$), $Q = \sum q_i$ must be balanced by the heat efflux through the cylinder generators, which is the cylinder surface, Fig. 8.1.

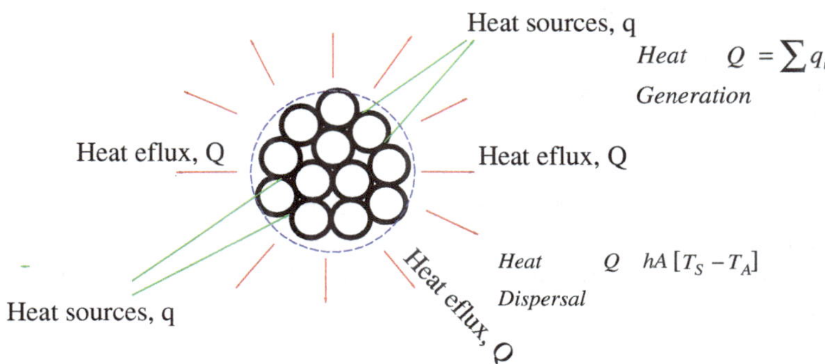


**Fig. 8.1** Schematic of the heat generation and dispersal

This is achieved by surface conduction, and the law that is conventionally used for forced convection from the cylinder surface is $Q = hA(T_S - T_A)$

where $T_A$ is the ambient temperature,

$T_S$ is the temperature on the surface of the rope
h is the surface conductivity, heat transfer coefficient
and A is the surface, in this case the circumference of the rope (=πD).

The following tables show approximate heat transfer coefficients and are reproduced from Mills [1]

| Medium | Heat transfer coefficient ($Wm^{-2}K^{-1}$) | Heat transfer rate ($W\ m^{-2}$) for $T_S = 40$ °C and $T_A = 15$ °C |
|---|---|---|
| Still air | 10 | 250 |
| Air at 5 $ms^{-1}$ | 50 | 1,250 |
| Water at 5EC | 1,000 | 25,000 |
| Water spray | 1,500 | 37,500 |

and from Eshbach [2]

| Order of magnitude of h | BTU/hr $ft^2$ °F | $Wm^{-2°}K^{-1}$ |
|---|---|---|
| Gases, natural convection | 0.9–5 | 5.157–28.65 |
| Flowing gases | 2–50 | 11.46–286.5 |
| Flowing liquids, non-metallic | 30–1,000 | 172–5,730 |

(*Conversion* 5.73*BTU $hr^{-1}\ ft^{-2}\ °F^{-1}$ gives $Wm^{-2}\ K^{-1}$ )

Although surface modelling has been quoted above, surface radiation is also an active mechanism for dispersal of heat to the ambient surrounding.

In this case, the heat efflux mechanism model is $Q = e\sigma A\left(T_S^4 - T_A^4\right)$

where $T_S$ is the surface temperature in °K (=°C + 273)

$T_A$ is the ambient temperature
E is the surface emissivity

= 0 for polished surface,
= 1 for matt black surface

Φ is a Stefan-Boltzmann constant (= 5.72 $10^{-8}Wm^{-2°}K^{-4}$)
A is surface area (= πD)

For a surface temperature of 40 °C and ambient temperature 15 °C, the heat radiated is 155 W $m^{-2}$, small compared with the still conduction rate. This has been included in the modelling and it's inclusion has modified the surface temperature by no more that 2 °C.

## References


1. Mills NJ (1993) Plastics, microstructure and engineering applications, pp 95 et seq., Edward Arnold, London
2. Eshbach OV, Souders M (1975) Handbook of engineering fundamentals, p 1113. Wiley, New York

# About the Author

The author graduated in Aeronautical Engineering from the Imperial College, London, and spent 10 years with the Canadian Defence Research Board (DRB) working on various aspects of hypersonic dynamics and high speed impact. In this time he also studied at Toronto University for an MEng and Ph.D., sponsored by the DRB. He then went to Bristol University to work on the dynamics and aerodynamics of containers suspended below helicopters.

The next position was at University of Manchester Institute of Science and Technology (UMIST) as a Lecturer and then as Reader; at UMIST he was initially involved with impact dynamics, applied to nets and body armour. This developed into liaison and collaboration with the Textile Department at UMIST and involvement into cable dynamics and ultimately ropes. He was awarded the DSc. (Applied Mechanics) by Imperial College.

He was one of the Directors involved in forming the company Tension Technology International Limited (TTI), and as a consultant was instrumental in undertaking various contracts in the modelling and assessment of rope mechanics and behaviour. He has since retired from UMIST and from TTI as a Director but is still active as a consultant.

C. M. Leech, *The Modelling and Analysis of the Mechanics of Ropes*,
Solid Mechanics and Its Applications 209, DOI: 10.1007/978-94-007-7841-2,

[illegible]

[illegible]

[illegible]

# Appendix

## Tentative Scaling Relations

The purpose of this section is to establish for engineering applications, very approximate and tentative laws between various geometrically similar structures but of different sizes.

This is based on the hypothetical comparison of various sized rope structures, diameter d, with like construction and materials. The operating condition is the same relative point, for example the terminal or breaking load or a nominal strain, and to the same operating geometry, i.e. for bending, the same relative curvature d/D.

- Loading
  The loads and contact forces are area proportional $\sim d^2$
  The slip measurement is size proportional $\sim d$
- Friction
  Linear contact force $\sim d^2$ Slip $\sim d$
  Scissor torque $\sim d^3$ Scissor angle is size independent
  Abrasion, damage/event $\sim d^3$
- Wear
  Wear criterion, area dependent $\sim d^2$
  Accumulated damage $\sim d$
  Life (cycles) $\sim d^{-1}$
  Thus the wear rate and heat generation is size cube dependent $\sim d^3$
  The wear criterion and wear area are probably area dependent $\sim d^2$
  This suggests that the accumulated damage is size dependent, $\sim d$ and consequently life, cycles to failure is inversely related to size, that is Life $\sim d^{-1}$
- Heat
  Heat dissipation, size dependent $\sim d$
  Temperature Differential, (Abrasion Rate/Dissipation Rate) $\sim d^2$
  Heat dissipation, being surface area dependent is size dependent $\sim d$ and consequently the temperature differential, surface temperature—ambient temperature, is area dependent, $\Delta T \sim d^2$.

C. M. Leech, *The Modelling and Analysis of the Mechanics of Ropes*,
Solid Mechanics and Its Applications 209, DOI: 10.1007/978-94-007-7841-2,

Zeitfracht Medien GmbH
Ferdinand-Jühlke-Straße 7
99095 Erfurt, Deutschland
produktsicherheit@kolibri360.de